A culture of curiosity

Manchester University Press

A culture of curiosity

Science in the eighteenth-century home

Leonie Hannan

Manchester University Press

Published by Manchester University Press
Oxford Road, Manchester M13 9PL

www.manchesteruniversitypress.co.uk

British Library Cataloguing-in-Publication Data
A catalogue record for this book is available from the British Library

ISBN 978 1 5261 5303 6 hardback

First published 2023

Typeset by
Deanta Global Publishing Services, Chennai, India

Contents

Figures

Acknowledgements

Taking place during multiple lockdowns and in an era of existential crisis, the process of writing this book has been fraught. Nonetheless, the strange influences of this time, both intensely domestic and devastatingly global, have made me think and rethink the home as a space of work and thought. Lacking the tacit knowledge of my book's subjects, I made very little serviceable sourdough, but I often contemplated the affective, intellectual and practical concerns engendered by the home, its temporal patterns and its regimes of labour. The book would not have materialised if it were not for the solid help and encouragement of many people.

Often beamed in from their own domestic spaces, friends and collaborators have been crucial to this project, in particular: Ananay Aguilar, Polly Bull, Gemma Carney, Helen Chatterjee, Sarah Longair, Catriona McKenzie, Kate Smith and Caroline Sumpter. I have had the good fortune to think with these people over many years. I have a long-term debt to Mat Paskins for helping me to see things in new and enlivening ways.

As a work of archival research, I am obliged to very many institutions and individuals but am particularly grateful for the kind support of Evelyn Watson at the Royal Society of Arts and also for Anton Howes' insights into this collection. Early on in the research, I was a visiting scholar at the Winterthur Museum, Garden and Library in Delaware, USA and benefitted from the expertise of their curators including the late and great Linda Eaton. As the manuscript neared completion, I was fortunate to secure a fellowship with the Descartes Center at the University of Utrecht, Netherlands and hugely valued the collegiality of Marjolijn Bol, Sven Dupré, Marieke Hendriksen, Grace Kim-Butler and Henrike Scholten. Over many years and in several jurisdictions, I have been fortunate to meet with generous opening hours and knowledgeable support in a multitude of archives. This book would not be one about Ireland if it weren't for the assistance of a historian of incredible talent, Ruth Thorpe, who foraged in archives when I could not.

Particular thanks go to those kind people who looked at drafts: Penelope Corfield, Kate Smith, Caroline Sumpter and the three anonymous readers. It has been a pleasure to work with the Manchester University Press team. I am also grateful to *Cultural and Social History* and Bloomsbury Academic for giving their permission to reproduce aspects of my work published in their volumes.

As with all large projects, this would not have been any fun without friends and family, especially Evi Chatzipanagiotidou, Becky Fishley, Julie Mathias, Sally Smith, Liza Thompson, the Hannans and my 'secret garden' neighbours. Very many trade unionists have made the increasingly troubled environment of higher education bearable, for the present and – I hope – the future. I am most grateful to Véronique Altglas for her friendship and comradeship. I wouldn't have embarked on this subject without long walks and interesting discussions with Joey O'Gorman.

Researched during a period when I began to wonder about parenthood and written after the fact, this book is dedicated to Aphra.

Abbreviations

BL	British Library
DCLA	Dublin City Library and Archive
DRO	Derbyshire Record Office
FHLD	Friends Historical Library Dublin
HRO	Hampshire Record Office
JRL	John Rylands Library
NLI	National Library of Ireland
PHA	Petworth House Archives
RSA	Royal Society of Arts
SA	Shropshire Archives
TCD	Trinity College Dublin
TNA	The National Archives (UK)
UM	The Ulster Museum
WCRO	Warwickshire County Record Office

Introduction: cultures of enquiry in the eighteenth-century British world

This is a book about the character of enquiry in the eighteenth century. It focuses on the years *c.* 1660 to 1830, an era synonymous with 'Enlightenment' and consolidation of the 'new science'. A vast body of scholarship has discussed the activities of this period's famous intellectuals: fellows of the Royal Society, university men, letter-writers in far-reaching networks of scholarly exchange or the new industrialists making connections between the art and science of manufacture. The individuals at the heart of this book are more difficult to categorise; their achievements were rarely proclaimed in print. They were members of a large and diverse population of the intellectually curious and they conducted their enquiries from home.

The eighteenth-century home was a complex space, capable of providing its inhabitants with sustenance of a physical, social and emotional kind. As such, it was also an environment uniquely conducive to scientific work. The materials, equipment and skills of home produced the goods necessary to feed, clothe and heal a family. Households were sustaining and generative; they were simultaneously the sites of childbirth and cheese-making. Many aspects of domestic labour demanded in-depth material knowledge and techniques were honed through repetition. As Bathsua Makin observed in 1673, 'To buy wooll and Flax, to die scarlet and purple requires skill in natural philosophy.'[1] Thus, people worked busily and skilfully to achieve the necessary cycles of production and consumption. They cheated the deprivation of winter with preserved fruit and meat and they carefully recorded the results of their resourcefulness in pounds, shilling and pence. These domestic practices, in all their variety, equipped occupants with the tools of their intellectual trade. Whether it was the secrets of nature, the physical properties of materials or the changing features of the night sky, individuals of a curious disposition used their domestic space, possessions and understanding to find out more. The extent to which these little-known domestic experimenters contributed to larger cultural and intellectual developments is a question worthy of an answer.

In offering an answer, this book strives to overcome systems of value that have promoted an exclusive conception of intellectual culture, thereby reassessing that culture. The research draws on disparate archival survivals, from inventories to life writing, that collectively reveal the scale of popular engagement with natural knowledge. This knowledge-making of the many was comprised of practices, communications and exchange, and was forged in the porous spaces of home. However, the discussion is less concerned with the 'knowledge' itself than with what people were doing, how they understood their activities and how their participation intersected with wider currents of thought. Far from scientific enquiry being a rarefied activity conducted by a special few, the chapters that follow reveal it as one integrated aspect of the variegated labour of home.

Of course, some people have seen their contributions to knowledge better documented than others. Though beyond the scope of this study, since the turn of the twenty-first century, research has engaged more concertedly with indigenous knowledge and the experimental activities of free and enslaved Africans in colonial contexts.[2] This scholarship begins to address long-term historiographical silences, which are themselves the result of unequal and unjust power relations. Another outworking of such biases is that women are less likely to be described as producers of knowledge than men, although a vibrant literature has emphasised their engaged consumption of ideas in this era.[3] The same could be said of many lower-status men.

Artisans are one category of (predominantly male) workers who have seen their contributions to science acknowledged.[4] By contrast, historians often view the labour of domestic servants as creatively unproductive. This book shifts focus from the labour enacted in the workshop to that of the home. In doing so, servants' detailed understanding of materials, techniques and technologies become visible.[5] By making plain the agency of marginalised people and the centrality of domestic labour to the more celebrated aspects of eighteenth-century cultural life, the characteristics of enquiry take on a different complexion.[6]

However, a focus on the material does not banish language to the sidelines. Taxonomy and terminology were crucial to many eighteenth-century scholars and experimenters, many of whom spent their time naming things, describing processes and identifying relationships between things. The process of naming things was, of course, materially embedded and Carl Linnaeus himself thought that plants could 'reliably speak their [own] character' and be named accordingly. Taking this belief seriously, 'the eloquent testimony of that epoch's material worlds' has become a key concern of many recent Enlightenment studies and is a central interest of this book.[7] However, it was not until the 2000s that scholarship fully acknowledged that some parts of the British world even experienced an 'Enlightenment'. For example, Ireland was traditionally cast as an intellectual backwater. Irish examples are drawn upon extensively here and this book's findings build on the growing body of work that demonstrates the cultural and scholarly vibrancy of this colonised island.[8]

In the chapters that follow, several categories of people are analysed, including servants, masters and mistresses; apprentices, tradespeople, professionals and the leisured, landed classes. All of them engaged in enquiry using practices learned at home. The research draws on a messy multiplicity of domestic, manuscript sources and contemporary print culture. Whilst the terms 'home' and 'household' are used interchangeably, an encompassing understanding of that space is employed, one that incorporates the land surrounding a property – outhouses and alike – as part of the infrastructure of domestic production. Some of the examples also show people using skills honed at home, outside of its parameters. The research thus focuses on domestic practices that could lead to increased understanding of natural phenomena. In doing so, it emphasises the experience of previously under-studied experimenters as opposed to the results of their enquiries.[9]

Of course, experience is a fraught category of historical analysis. Nonetheless, since E. P. Thompson argued that the emergence of the working class at the turn of the nineteenth century rested on the experiences of working people, social, economic and cultural historians have all engaged with this realm.[10] Even fragmentary evidence of the mundane or routine actions of everyday existence offers useful information about people's common expectations, limitations and aspirations.[11] To look up, from below, offers new insight – it aims to describe the larger sphere; it is also difficult to achieve and tends to require more methodological justification.

The rise of cultural history and the consolidation of gender history as a field at the end of the twentieth century have ensured the place of marginalised groups and their experiences at the centre of historical understandings of the past.[12] Whilst historians of culture and gender have often amplified personal experience and subjectivities, they have also articulated how individual experience was shaped by larger structures and power relationships. In these ways, several generations of scholars have paid attention to the quotidian with the precise intention of subverting the societal hierarchies that obscure the agency and intention of 'ordinary' people. This book continues that tradition.

In the same vein, historians of science have rapidly advanced research into the social contexts, practices, objects, environments and embodied experiences of scientific enquiry in a process described by Steven Shapin as 'lowering the tone'.[13] This development responded to the critique that histories of science were much more likely to examine the product of knowledge-making than the process itself, despite the significance of the process for the outcome.[14] These efforts to contextualise intellectual activity have resulted in a more material and embodied understanding of enquiry, as historians of science have argued – ideas simply cannot travel without objects and bodies.[15] With this in mind, global histories of science have championed the agency that exists at all social levels and emphasised the links in the chain, the human and non-human interactions, the intermediaries and brokers.[16] Scholarship has also revealed deep-rooted practices of repair and reuse of both domestic and scientific equipment alongside the novel possibilities offered by technological advancement, thereby unsettling traditional interpretations of 'Enlightenment science'.[17] This book contributes to a developing social history of domestic space, material culture and science that can move firmly beyond a model that confers the lion's share of agency on the male and the genteel.

Histories of science are most interested in experience that is 'contrived and disciplined', the kind that conditions the body and mind in specific ways.[18] Forms of experience such as observation or experimentation could be converted into verifiable knowledge, depending on the context in which they occurred and the status of the actors involved. Despite seventeenth-century criticisms of Aristotelian natural philosophy as overly focused on wordy reasoning as opposed to observable phenomena, the connection between knowledge and the senses was an ancient one.[19] Whilst the complexity of the developments described as the 'new science' is often underplayed, the changes they wrought did increase emphasis on sensory experience as a route to knowledge.[20]

Whilst informed by these histories, this book takes a different approach. The evidence analysed offers insight into experience of this kind – focused and disciplined – but also of the kind discussed above, everyday and ordinary. In fact, this research sees the former as an outgrowth of the latter. The practice of scientific knowledge-making is understood as embedded in the conditions, labour and knowledge associated with domestic life. This shift in perspective allows a different range of scientists to move into the foreground.

Women have often been understood as appendages to scientific enquiry, rather than central players with questions and practices of their own. Their experiences of scientific enquiry were, of course, conditioned by their circumstances and the gendered prescriptions of the day. That said, a range of scholarship has illuminated the important role many women played in the scientific knowledge-making in this period and others.[21] Lynette Hunter and Sarah Hutton's foundational contribution, *Women, science and medicine 1500–1700*, identified the kitchen and stillroom as female spaces of key importance to chemical, biological and medical knowledge.[22] Their analysis saw 'modernity' as the catalyst for ultimately separating knowledge from the domestic. Whilst the findings discussed in this book deviate from that model of change over time, Hunter and Hutton's proposal that knowledge was embedded in the everyday rightly endures. This book makes that case for the eighteenth century.

With the diversification of the history of intellectual work comes a tendency to see knowledge as circulating rather than as being produced neatly in one time and place and by one person.[23] In the chapters that follow, a wide range of curious individuals are examined, people who used their own experience to develop natural knowledge and who often regarded their personal experience as scientifically valid. As Carolyn Steedman has noted, analytical advantage is enhanced by comparing the everyday experiences of different kinds of people.[24] This study considers women as well as men, servants as well as employers and a spectrum of households from the modest to the elite. By analysing the marginalised alongside the privileged, a more accurate understanding of both is possible.

Curiosity

Curiosity was rife in the eighteenth century, and many acted upon it to investigate the natural world. These enquiries of different types and scales amounted to more than the sum of their parts. Taken together, the actions of many interested individuals can be considered a culture of curiosity, with ramifications for the characterisation of intellectual life in this period. Whilst this study focuses very strongly on the particulars of practice – what people were actually doing – it does not ignore the motivation for such actions. In a domestic setting, practices of trying and testing in order to alter and finesse material processes were commonplace. No clear distinction exists between this bedrock of home oeconomy and activities of a more investigative quality. After all, the development of a fermentation process might secure a better consumable product as well as elicit an understanding of the properties of ingredients, singly and combined, and their reactions under a range of conditions. The urge to take a domestic observation or experiment a step further than strictly required was fuelled by curiosity and this is a recognisable trait across a wide range of examples discussed here.

Whilst a lively 'culture of curiosity' has been well documented by scholars of eighteenth-century Britain, many definitions of this phenomenon remain exclusive. Typically the activities of *curiosi* and *virtuoisi* sit centre stage – in other words, landed and educated men, motivated by awe and articulate in their wonder.[25] However, even in the hands of the wealthy, curiosity was not always considered a force for good. Where there were the curious, there was self-indulgence, unfettered desire and a dangerous rejection of the status quo.

When curiosity fuelled the collecting of specimens or artefacts, passion, avarice and an urge to possess and control might follow.[26] The loss of discipline associated with curiosity was thought particularly acute for women and the lower classes whose intellectual pursuits were more likely to be censured.

The negative connotations of curiosity have been enduring. In an anonymous interview published in *Le Monde* in 1980, a philosopher – later revealed to be Michel Foucault – commented, 'Curiosity is a vice that has been stigmatized in turn by Christianity, by philosophy, and even by a certain conception of science.' Curiosity had become equated with futility.[27] Nevertheless, in the late twentieth century, Foucault dreamt of 'a new age of curiosity' based on the generative possibilities of this urge:

> it evokes 'concern'; it evokes the care one takes for what exists and could exist; a readiness to find strange and singular what surrounds us; a certain relentlessness to break up our familiarities and to regard otherwise the same things; a fervour to grasp what is happening and what passes; a casualness in regard to the traditional hierarchies of the important and the essential.[28]

The historical examples discussed here strongly reflect this characterisation of curiosity. A motivation to grasp the particularities of the familiar whilst attending to the unfamiliar is visible in the domestic enquiries of many. By exploring the conditions and substance of their activities, the significance of seemingly unimportant people, things and actions is reappraised.

Historians have identified a new drive for innovation as motivating social and economic change in the eighteenth century. Recent scholarship has described this 'improving mentality' as contagious; in other words, ardent individuals were keen to share their intention to innovate and ensured that innovation was a practice that spread, person to person.[29] This assessment maps onto a general upsurge in popular engagements with science. An inclusive culture of innovation, dependent more on personal motivation than role, education or training, corresponds closely with the findings of this book. However, for the individuals explored here, an urge to innovate appears as just one strand of many motivations – which, taken as a whole, are better described by the term curiosity. In fact, this study illuminates the compound nature of the impulse to enquire.

For some, knowledge for its own sake was inducement enough. For others, an interest in uncovering the 'secrets of Nature' in order to contribute to a larger, collective project of knowledge-making was the motivation. Concern with the economic and cultural vitality of the nation or the furthering of Britain's global, imperialist exploits also drove enquiry. These external factors combined with personal agendas, such as the discharge of a given role (or moving beyond its boundaries) or the performance of an identity (or the subversion of expected norms).

For the curious enquirers of this volume, exchange and community were often crucial. That is not to say that the lone wolf of home experiment did not exist, but the people discussed in detail here sought out communication with others on their subjects of interest. This volubility might be an artefact of the archive – those who wrote a lot were more likely to have their words preserved to the advantage of the historian. However, it is also true to say that a network of like-minded friends eased a range of obstacles to enquiry, especially for the intellectually marginalised – information, ideas, reading material, specimens and access to other people all came to those who engaged with a network of contacts on matters of mutual interest. The curious individuals who populate this book did not all understand themselves to be a part of a national or international culture of curiosity *per se*, but they typically believed in a community of enquiry beyond the threshold of their own home. This recognition of shared purpose acted to affirm curiosity itself.

Making and sharing knowledge

Access to information

Eighteenth-century people learned new knowledge and skills in many different ways, much as people do today. For good reasons, histories of knowledge have tended to emphasise the tools and technologies at humankind's disposal. The invention of the printing press, the consequent proliferation of print culture, the growth of certain institutions, the emergence of new spaces for public debate and the increasing availability of diverse material culture have all been identified as drivers of new ideas in this period. This approach has prioritised certain technologies and cultural practices as routes to making knowledge over others, namely 'books and bookishness' and the design and production of scientific instruments.[30] Whilst it is undoubtable that access to a broader range of reading material and the use of specially designed apparatus helped curious individuals expand their understanding of the world, other practices did so too. Forms of labour, observations, conversations with like-minded others, practices of record-keeping and habits of collecting and categorisation were all valid modes of learning that provided paths to new understanding. All of these actions took place at home.

For those for whom the price of a book was too much, a wealth of other reading material was accessible. Magazines and periodicals also proliferated in this period and many of these publications covered subjects of a scientific nature. It is also worth remembering that the well-to-do and wealthy were avid consumers of cheap print alongside those whose means extended no further. For younger audiences, a growing educational literature presented natural knowledge in accessible ways – often utilising the format of a conversation to inform the child reader.[31] Didactic formats such as dialogues, grammars and lexicons shaped the way people learned, and the gendered prescriptions in published texts went through considerable transformation in this period, reflecting changing social attitudes towards education and masculinity.[32]

Here, the development of knowledge is seen as widely distributed – both in terms of the kinds of people and things involved in enquiry and also concerning the manner in which learning came about. Eighteenth-century people learned through reading and doing and engaged collaboratively with other people, objects and spaces to discover the answers to questions of all kinds.[33] Actions that were understood in the context of 'keeping house' also fulfilled investigative goals. Tacit knowledge learned in the process of brewing ale was put to work in the service of curiosity.

Networks of exchange

Eighteenth-century society supported a growing population of letter-writers who were able to forge geographically extensive and socially encompassing networks. Whilst correspondence oiled the wheels of all manner of human relationships, from courtship to commerce, intellectual life was a major beneficiary of the post coach. Networks of a modest scale were facilitated through correspondence just as much as person-to-person contact, and many exchanges involved both. In urban environments, it was more likely that contacts might be reachable on foot in nearby streets, but a letter could overcome considerable distances to reach the rural or remote.[34] Members of the landed classes often enjoyed greater mobility than working people, the wealthiest benefitting from the seasonal migration between city and countryside. However, neighbourhood connections were powerful across the social hierarchy. So, whilst the question of scale is an interesting one, the notion of contributing to a collective endeavour of some kind was commonly held, even if that sense of community had to be derived primarily from the periodical press.

The contributions of the 'big names' of early modern science are now understood in the context of their diligently fostered and far-reaching epistolary connections.[35] Experimental work involved multiple actors and, to gain traction, a new finding needed the validation of a wider community.[36] This community encompassed a wide social makeup. The natural philosopher and general polymath, Robert Hooke (1635–1703), was regularly in personal contact with labourers, servants, craftsmen (e.g. glassblowers or clockmakers), gentlemen and noblemen as he conducted his many projects across the city of London.[37] As historians have looked back on the work of eighteenth-century natural philosophers, the influence of the Romantic notion of authorship is clear. A scholarly obsession with attribution to a single (usually) male instigator of a particular idea or work has been difficult to dislodge and has served to obscure important characters and characteristics of knowledge-making.[38] The individuals in this book might have had fleeting contact, at best, with institutions of intellectual note, but they were often well networked and entirely capable of using their contacts to further their interests.

Aside from more informal networks of letter-writers, institutions also played an important role in connecting people and providing a destination for information garnered in other environments. This period saw the rise (and sometimes fall) of a number of learned societies that have been considered influential in shaping intellectual life.

With their different emphases, the Society of Antiquaries, the Royal Society, the Dublin Society, the Society for the Encouragement of Arts, Manufactures and Commerce, the Physico-Historical Society and a plethora of local philosophical societies all accepted incoming correspondence from scattered individuals and an outlet in print for the findings that emerged from those diverse quarters.[39] On a smaller scale, clubs and domestic sociability offered other meeting points for the intellectually engaged.[40]

The eighteenth century famously witnessed a flourishing of debate in the coffeehouses of urban centres. These spaces were public although it is worth remembering that the demonstrations and discussions that took place there were not universally accessible.[41] Coffeehouse culture was intimately connected with the flourishing of periodical publications, such as the *Spectator* or the *Gentleman's Magazine*. They were also associated with the Royal Society's networks of scientific sociability.[42] Moreover, coffeehouses performed a range of services for publishers, including advertising and collecting book subscriptions.[43] These spaces offered an alternative social space, one that could accommodate debate, whilst avoiding the seriousness of more formal locations. The sheer regularity of some individuals' visits to coffeehouses is witnessed by the many letters addressed to these premises instead of homes or offices. Whilst coffeehouse culture has received much scholarly attention, it seems likely that other social spaces such as inns and taverns also provided opportunities for sharing news and ideas.[44] However, whilst these semi-public spaces proliferated in this period and provided some people with new opportunities to learn and share ideas, they excluded many others. Here, the household is the key unit of analysis, not to diminish the importance of other spaces and places but because the intellectual possibilities of the home were significant and have been, by comparison, under-explored. Moreover, it is through shifting focus from talking and exchanging to doing and making that the importance of the home as a site of knowledge-making becomes clear.

A challenge of locating scientific activity in one kind of space is the difficulty in simultaneously tracking its leaps and bounds between people, through objects and across boundaries. However, if knowledge-making is treated as a communicative activity in itself, then the unhelpful distinctions between the making and disseminating of knowledge can be diminished, distinctions that often locate the making with the privileged and imagine the communication of that knowledge as eventually reaching the marginal.[45] One way to access the talkative dimension of domestic knowledge-making is to focus attention not on the print culture of this period, but on the manuscript materials of everyday life – recipe books, account books, inventories, lists and receipts. This is the approach taken here. Among these manuscript survivals, it is possible to imagine the curious taking pen in hand, producing as well as consuming lines on the page.

Working at home

There are distinct advantages to using the home to examine larger currents of intellectual life. For one, domestic space was available to almost everyone and had some unifying characteristics, such as the inclusion of spaces and equipment to aid in provisioning, socialising and resting, although homes differed dramatically in their scale and affordances. The household and its inhabitants also sat in a conceptual relationship with the state and whilst it serves the purposes of this research to look at the home as a site of intellectual work, those engaged in these activities were likely to have understood the home as a miniature nation.[46] Seeing parallels between domestic order and national prosperity, many eighteenth-century householders also made connections between their own home experiments and the pressing concerns of their age.

The evidence for this book takes in modest urban dwellings above shops through to grand country seats, set in acres of parkland. It draws on people who lived, mainly, in the British Isles including Ireland, but it sometimes looks to the emigrants who headed across the Atlantic to the east coast of America for information about the kinds of homes they built and practices they undertook as they forged their lives in a very different climate and terrain.[47] The somewhat blurred edge of the group of examples used here allows the book to look outwards to the truly global circuits of trade, networks of exchange and colonial relations of domination and extraction whilst retaining its centre of gravity in the domestic spaces of British and Irish people across these islands. Undoubtedly, increasing numbers of people read about and even travelled to places that would have seemed highly exotic – almost mythical – to previous generations. As botanic gardens, some long-established, gathered seeds from the Caribbean, the East Indies and Australia to cultivate in British soil, paradoxically, the world seemed increasingly knowable and exponentially variegated in its natural wonder.

Household labour, space, materials and things

Domestic work was constant and larger households contained one or more rooms that were designed and equipped to produce domestic necessities. There was much 'doing' at stake in eighteenth-century homes. At this time, the home made many more of its everyday necessities from raw ingredients than is common in twenty-first-century western society. Moreover, where a household had access to land, raw ingredients might be cultivated. In the absence of refrigeration, preserving and pickling were strategies for making food last and ensuring a varied diet during the cold winter months. Household accounts and recipe books reveal the wide range of ingredients familiar to those in charge of household provisioning, but also the expansive repertoire of processes enacted on those ingredients in order to maximise their value and use.[48]

Domestic work has often been treated as an unchanging continuum, in contrast with the large-scale changes that are seen to take place in other forms of work in this period. Moreover, shifts in labour relations in the eighteenth and early nineteenth centuries are often cited as key determinants of the emergence of 'modernity'. The way work is understood, then, is of critical importance to understanding this period as a crucible of social and economic change. However, as Jane Whittle has eloquently unpacked, common scholarly definitions of work are flawed, principally because they misunderstand work that took place in the home, especially when it was done by women.[49] In this context, it is perhaps unsurprising that a significant body of scholarship on the home as a place of work focuses on the labour relations of domestic service, precisely because of the insight they offer into the larger socio-economic shifts associated with industrialisation.[50] By contrast, Carolyn Steedman's analysis casts new light on the qualities of servants' labour, the feelings it provoked and the material dimension of domestic service, arguing that 'material things – jokes, jests and the well-set jam a maidservant had just produced – were objects and entities, part of the social world' and as such critical to understandings of the 'social order' of that time and place.[51] It is worth emphasising that not all domestic labour promoted scientific enquiry, and that the eighteenth-century home entailed a great deal of drudgery. Nevertheless, small acts of domestic labour were meaningful in manifold ways, and certainly no less so than the flick of an official's pen.

Twenty-first-century studies of eighteenth-century domestic work have examined practices relating to food, record-keeping and domestic upkeep and revealed their larger social relevance.[52] This scholarship builds on at least fifty years of research by gender historians that has debated women's experiences of domestic life and labour at length, also addressing male contributions to domestic work.[53] Such analysis of gender, work and the home has been closely shadowed by debates concerning 'public' and 'private' space. The line between the two has increasingly been viewed by historians as blurred and a stark dichotomy is now largely rejected. The research presented here aims to further dismantle these restrictive categories, which have frequently served to obscure the action taking place around, outside or in contradiction with them.

In approaching the subject of domestic work, this book understands the pre-industrial economy as one that included lots of unpaid work that was often geared towards subsistence. The chapters that follow see no clear distinction between domestic processes of money-earning and money-saving and incorporate men and women alike in the labour of the home. It is hoped that by avoiding unhelpful contortions that count male domestic work as 'productive' and female domestic work as 'unproductive', this book can contribute to a more accurate and heterogeneous understanding of the work of the home, extending that definition further to include activities of enquiry.[54]

In domestic tasks, individuals derived status, performed gendered roles and cared for others. When gentlewoman Anne Dormer of Rousham House in Oxfordshire complained that her husband was 'much taken with all sorts of cookery and spends all his ingenuity in finding out the most comodious way of frying broileing resting stewing and preserving his whole studdy'[55] or 'loiter[ing] aboute, somtimes stues prunes, somtimes makes Chocolate, and this somer he is much taken with preserving',[56] her point was clear – these tasks and who did them mattered, not only for the home economy but also for the moral order of her household. These kinds of domestic experiments are the subject of this book. Whilst conduct manuals of the period were clear in their prescriptions for how domestic work should be undertaken, for what purpose and by whom, plenty of individuals ploughed their own furrow. A permeability existed between skills learned at home and those put into practice in other places, whether they were institutional, artisanal, industrial or commercial. Moreover, for many, the home was also the workshop.[57]

A home for 'Enlightenment science'

As Steven Shapin highlighted in the 1980s, domestic space was commandeered and adapted by natural philosophers to serve their investigative needs.[58] These men employed common household utensils, furniture and spaces to serve experimental ends, and employed materials ready to hand in the home to learn about nature.[59] The motives for such practices were varied, ranging from practical and economic constraints of poverty and scarcity to religious and social values of thrift and stewardship.[60]

Whilst historians have demonstrated that the household, including gardens, were crucial spaces in 'the making of modern science', the ways in which different domestic activities interacted are less well understood.[61] Histories reflect the differentials in status attributed to the process of making jam versus conducting a scientific experiment. However, both of these activities involved an in-depth knowledge of material properties, the use of specialised equipment, the heating and cooling of materials to change their quality and the tacit knowledge of having performed these actions repeatedly and with particular aims in mind. Moreover, scholarship can sometimes compound this distinction by failing to recognise the high status of some domestic labour – especially the kinds of knowledge and skill that were required to operate a stillroom effectively.

A number of studies have traversed this terrain and in doing so situated recipe books at the centre of day-to-day investigations of natural knowledge in the genteel household, positioned cooking as an epistemological practice and identified the kitchen as an explicitly experimental space.[62] This book demonstrates the way skill, tacit knowledge, technology and the rhythm of daily work in the home created the conditions and aptitudes for scientific enquiry. It does so by examining what people were doing at home and discovering the wider significance of these practices for eighteenth-century science.

Domestic practice as a route to enquiry

In this book, the discussion of practices takes up a good deal of space. This term is helpful because it captures something of the grey area between the characteristics of an action and the understanding it confers. In Alan Warde's discussion of philosophical accounts of practice, he distinguishes between 'praxis' as the whole of human action and 'practice' as routinised behaviours involving combinations of bodily, mental and emotional activity in the context of existing knowledge or know-how.[63] These practices are, by definition, 'social'.[64] Scholarly understandings of domestic practice have been influenced in particular by the theoretical interventions of Pierre Bourdieu and Michel de Certeau on practice and habitus and everyday practice respectively.[65] However, methodologically, it has been ethnographic approaches that have been most effective in accessing the dynamics of the home. Attending to these questions in a historical context is more challenging given the inability to observe practice in action. Nonetheless, by carefully considering extant material, spatial and textual remnants, it is possible to reconstruct elements of quotidian domestic practice.[66] The approach taken in researching this book represents a resourceful use of a variety of genres of evidence and makes the case for working across these categories to access the everyday activity of the home.

The kinds of knowledge or understanding explored in this volume include self-consciously scholarly activity alongside pursuits that occupy a more marginal space in histories of intellectual life. Prominent are activities that combine accumulated material understanding and refined technique, often described as 'tacit knowledge', in contrast to knowledge accrued primarily from text or developed in the Academy. Clearly, there has been a hierarchy of types of knowledge that has not always attributed much value to this kind of 'know-how'. As Michael Polanyi famously observed, 'we can know more than we can tell'.[67] A twentieth-century scientist turned philosopher, Polanyi was struck by the fact that humans could recognise a face amongst many thousands of other similar faces, without being able to describe with any degree of specificity its features. In this example, he recognised the importance of tacit knowing and argued that it played a central role in the development of scientific knowledge.

Historians from different fields agree that this was an era in which practical and sensory forms of knowledge assumed a much greater status.[68] Some studies articulate with great precision the central importance of haptic knowing in scientific developments.[69] Knowledge learned by doing was not only important for elite forms of experimental science, but it was also the route to understanding many different things for many different people. Far from being the poor man's laboratory, the home could afford a versatile space for the curious – whoever they may be. More than that, the home trained its inhabitants in skills and knowledge that they could very well put to the service of science.

Structure of the book

This book contains seven chapters organised into three sections. The chapters in Part I offer contextual information about the domestic environment, its spatial and material affordances and the record-keeping that underpinned home 'oeconomy' and enquiry. Part II focuses on a discussion of three household practices that promoted knowledge-making and the chapters in Part III consider the wider ramifications of these findings. The first chapter takes a close look at the materials that circulated through the early modern household and the spaces and equipment that allowed householders to develop material knowledge. It includes an examination of larger-scale shifts in room use in this period, in different regional contexts, along-side a discussion of the material culture of specific working rooms. Chapter 2 considers examples of the specialised skills developed at home by examining the way tacit knowledge, technique and practices of record-keeping operated in domestic environments. Chapter 3 is the first of three chapters that analyse a single domestic practice, in this case collecting. It explores the way that curious individuals used the acquisition of artefacts and specimens as an aid to learning and the networks of exchange that fuelled this process. In Chapter 4, the book moves to consider the explicitly exploratory activities of householders. Through a case study of two Dublin apprentices with an interest in astronomy, it focuses on the practice of observation. The following chapter shifts to the subject of experiments, exploring the world of British and Irish women silkworm breeders. Chapter 6 opens up the discussion of practices to consider the way people constructed their own intellectual authority and the relationship between personal activity and identity in eighteenth-century society and culture. The last chapter contemplates the larger questions of influence in eighteenth-century intellectual life, aiming to re-consider the culture of enquiry based on the findings of this study.

Taken together, these chapters argue that the environment, personnel, materials and activities of the home provided the conditions for scientific practice in this period. In doing so, this book uncovers a large population of curious enquirers in eighteenth-century society and acknowledges their role in the discovery of nature's secrets.

Notes

1 Bathsua Makin, *An essay to revive the ancient education of gentlewomen* (London, 1673), as quoted in Lynette Hunter and Sarah Hutton (eds), *Women, science and medicine, 1500–1700: Mothers and sisters of the Royal Society* (Stroud: Sutton Publishing Limited, 1997), p. 3.

2 See, for example, Londa Schiebinger, *Secret cures of slaves: People, plants, and medicine in the eighteenth-century Atlantic world* (Stanford, CA: Stanford University Press, 2017); James Delbourgo and Nicholas Dew (eds), *Science and empire in the Atlantic world* (London: Routledge, 2008); James H. Sweet, 'Mutual misunderstandings: gesture, gender, and healing in the African Portuguese world', *Past and Present*, 203:suppl. 4 (2009), pp. 128–43.

3 See, for example, Jacqueline Pearson, *Women's reading in Britain, 1750–1835: A dangerous recreation* (Cambridge: Cambridge University Press, 1999); Margaret Spufford, *Small books and pleasant histories: Popular fiction and its readership in seventeenth-century England* (Cambridge: Cambridge University Press, 1981); Norma Clarke, *The rise and fall of the woman of letters* (London: Pimlico, 2004); Sarah Knott and Barbara Taylor (eds), *Women, gender and enlightenment* (Basingstoke: Palgrave Macmillan, 2005); Helen Berry, *Gender, society*

and print culture in late Stuart England: The cultural world of the Athenian Mercury (Aldershot: Ashgate, 2003); Margaret W. Ferguson, *Dido's daughters: Literacy, gender, and empire in early modern England and France* (London: University of Chicago Press, 2003); Hunter and Hutton, *Women, science and medicine*; Deirdre Raftery, *Women and learning in English writing, 1600–1900* (Dublin: Four Courts, 1997); Evelyn Arizpe and Morag Styles with Shirley Brice Heath, *Reading lessons from the eighteenth century: Mothers, children and texts* (Shenstone: Pied Piper, 2006); Berenice A. Carroll, 'The politics of "originality": Women and the class system of the intellect', *Journal of Women's History*, 2:2 (1990), pp. 136–63.

4 Pamela H. Smith, *The body of the artisan: Art and experience in the scientific revolution* (Chicago, IL: University of Chicago Press, 2004); Pamela H. Smith, Amy R. W. Meyers and Harold J. Cook (eds), *Ways of making and knowing: The material culture of empirical knowledge, 1400–1850* (Ann Arbor, MI: University of Michigan Press, 2014); Ann Secord, 'Science in the pub: Artisan botanists in early nineteenth-century Lancashire', *History of Science*, 32 (1994), pp. 269–315.

5 Carolyn Steedman, *Labours lost: Domestic service and the making of modern England* (Cambridge: Cambridge University Press, 2009).

6 See Christine von Oertzen, Maria Rentetzi and Elizabeth S. Watkins, 'Finding science in surprising places: Gender and the geography of scientific knowledge. Introduction to "Beyond the academy: Histories of gender and knowledge"', *Centaurus*, 55 (2013), p. 74 (pp. 73–80).

7 Adriana Craciun and Simon Schaffer, *The material cultures of enlightenment arts and sciences* (London: Palgrave Macmillan, 2016), p. 3; although it is worth remembering that Linnaeus's system deliberately overlaid and erased indigenous names.

8 James Livesey, *Civil society and empire: Ireland and Scotland in the eighteenth-century Atlantic world* (London: Yale University Press, 2009); Ian McBride, 'The edge of enlightenment: Ireland and Scotland in the eighteenth century', *Modern Intellectual History*, 10:1 (2013), pp. 135–51; Michael Brown, *The Irish Enlightenment* (Cambridge, MA: Harvard University Press, 2016).

9 This work takes inspiration from sociology of scientific knowledge and the Strong Programme's understanding of the importance of social factors in all scientific activity; see, for example, Finn Colin, *Science studies as naturalized philosophy* (Dordrecht: Springer, 2011), esp. chapter 3 'David Bloor and the Strong Programme', pp. 35–62.

10 Edward P. Thompson, *The making of the English working class* (London: Victor Gollancz, 1963); in terms of recent historiographical developments, Carolyn Steedman's defence of both experience and

the everyday as freighted yet revelatory concepts informs this book's research, see *An everyday life of the English working class: Work, self and sociability in the early nineteenth century* (Cambridge: Cambridge University Press, 2013), esp. pp. 25–7.

11 See, for example, Beverly Lemire, *The business of everyday life: Gender, practice and social politics in England, c. 1600–1900* (Manchester: Manchester University Press, 2012).

12 Joan Wallach Scott, 'Gender: a useful category of historical analysis', *The American Historical Review*, 91:5 (1986), pp. 1053–75; Hannah Barker and Elaine Chalus (eds), *Gender in eighteenth-century England: Roles, representations, and responsibilities* (New York: Longman, 1997); Mary O'Dowd, *A history of women in Ireland, 1500–1800* (Harlow: Longman, 2005); Krassimira Daskalova, Mary O'Dowd and Daniela Koleva, 'Introduction', *Women's History Review*, special issue: Gender and the cultural production of knowledge, 20:4 (2011), pp. 487–9.

13 Steven Shapin, *Never pure: Historical studies of science as if it was produced by people with bodies, situated in time, space, culture, and society and struggling for credibility and authority* (Baltimore, MD: Johns Hopkins University Press, 2010); see also Bruno Latour's seminal work, *Science in action: How to follow scientists and engineers through society* (Cambridge, MA: Harvard University Press, 1987); work by historical geographers such as David Livingstone has also had its impact, ensuring that the study of science attends to location, place and space: *Putting science in its place: Geographies of scientific knowledge* (Chicago, IL: University of Chicago Press, 2003).

14 Pamela H. Smith and Benjamin Schmidt (eds), *Making knowledge in early modern Europe: Practices, objects, and texts, 1400–1800* (Chicago, IL: Chicago University Press, 2007), p. 3; Smith, *Body of the artisan*.

15 See, for example, Sven Dupré and Christoph Herbert Lüthy (eds), *Silent messengers: The circulation of material objects of knowledge in the early modern Low Countries* (Berlin: LIT Verlag, 2011); Simon Werrett, *Thrifty science: Making the most of materials in the history of experiment* (Chicago, IL: Chicago University Press, 2019); Smith, *Body of the artisan*.

16 See, for example, Sarah Easterby-Smith, 'Recalcitrant seeds: material culture and the global history of science', *Past and Present*, supplement 14 (2019), pp. 215–42; James Delbourgo, Kapil Raj, Lissa Roberts and Simon Schaffer (eds), *The brokered world: Go-betweens and global intelligence, 1770–1820* (Sagamore Beach, MA: Science History Publications, 2009).

17 Werrett, *Thrifty science*; see also David Edgerton, *The shock of the old: Technology and global history since 1900* (London: Profile, 2008).
18 Lorraine Daston and Elizabeth Lunbeck (eds), *Histories of scientific observation* (Chicago, IL: Chicago University Press, 2011), p. 3.
19 Peter Dear, 'The meanings of experience' in Katharine Park and Lorraine Daston (eds), *The Cambridge history of science*, vol. 3 (Cambridge: Cambridge University Press, 2006), pp. 106–8 (pp. 106–31).
20 Lorraine Daston and Katharine Park, *Wonders and the order of nature, 1150–1750* (New York: Zone Books, 1998), p. 330.
21 See, for example, Ludmilla Jordanova, 'Gender and the historiography of science', *The British Journal of the History of Science*, 26:4 (1993), pp. 469–83; Patricia Fara, *Pandora's breeches: Women, science and power in the Enlightenment* (London: Pimlico, 2004); Michelle DiMeo, '"Such a sister became such a brother": Lady Ranelagh's influence on Robert Boyle', *Intellectual History Review*, 25:1 (2015), pp. 21–36; Ruth Watts, *Women in science: A social and cultural history* (Abingdon: Routledge, 2007); Patricia Phillips, *The scientific lady: A social history of women's scientific interests, 1520–1918* (London: Weidenfeld and Nicolson, 1990); and von Oertzen et al., 'Finding science in surprising places'; Katherine Allen, 'Hobby and craft: Distilling household medicine in eighteenth-century England', *Early Modern Women: An Interdisciplinary Journal*, 11:1 (2016), pp. 90–114.
22 Hunter and Hutton, *Women, science and medicine*, p. xii.
23 Or 'in transit' as James A. Secord has suggested; 'Knowledge in transit', *Isis*, 95:4 (2004), pp. 654–72.
24 Steedman, *Everyday life*, pp. 14–15.
25 Katie Whittaker, 'The culture of curiosity' in N. Jardine, J. A. Secord and E. C. Spary (eds), *Cultures of natural history* (Cambridge: Cambridge University Press, 1996), pp. 75–90; for an examination of cultural representations of curiosity see Barbara M. Benedict, *Curiosity: A cultural history of early modern inquiry* (Chicago, IL: University of Chicago Press, 2001); for scholarship that values local studies into early modern curiosity, see Neil Kennedy, *Curiosity in early modern Europe: World histories* (Wiesbaden: Harrassowitz, 1998) and R. J. W. Evans and Alexander Marr (eds), *Curiosity and wonder from the Renaissance to the Enlightenment* (Aldershot: Ashgate, 2006).
26 Krzysztof Pomian, *Collectors and curiosities: Paris and Venice, 1500–1800*, trans. Elizabeth Wiles-Portier (Cambridge: Polity, 1990); Stacey Sloboda, 'Displaying materials: Porcelain and natural history in the Duchess of Portland's museum', *Eighteenth-Century Studies*, 43:4 (2010), pp. 460–1 (pp. 455–72); Lorraine Daston and Katharine Park have argued that curiosity became less associated with lust after 1750

but they are referring to known natural philosophers only, *Wonders*, pp. 303–28.

27 Michel Foucault as quoted in Brian Dillon and Marina Warner, *Curiosity: Art and the pleasures of knowing* (London: Hayward Publishing, 2013), p. 22.

28 *Ibid.*

29 Anton Howes, 'The relevance of skills to innovation during the British Industrial Revolution, 1547–1851', working paper (2017): www.antonhowes.com/uploads/2/1/0/8/21082490/howes_innovator_skills_working_paper_may_2017.pdf (accessed 17 June 2021).

30 Smith and Schmidt, *Making knowledge*, pp. 2–3, see also Elizabeth Yale on the importance of seeing print and scribal cultures as working in tandem: 'Marginalia, commonplaces, and correspondence: Scribal exchange in early modern science', *Studies in History and Philosophy of Biological and Biomedical Sciences*, 42 (2011), pp. 193–202.

31 Michèle Cohen, '"To think, to compare, to combine, to methodise": Notes towards rethinking girls' education in the eighteenth century' in Knott and Taylor, *Women, gender and enlightenment*, pp. 224–42; see also Michèle Cohen, 'Gender and the public private debate on education in the long eighteenth century' in Richard Aldrich (ed.), *Public or private education? Lessons from history* (London: Routledge, 2004), pp. 15–35; and Natasha Glaisyer and Sara Pennell (eds), *Didactic literature in England, 1500–1800* (London: Routledge, 2017).

32 Michèle Cohen, 'French conversation of "glittering gibberish"? Learning French in eighteenth-century England' in Glaisyer and Pennell, *Didactic literature*, pp. 99–117.

33 See Edwin Hutchins, *Cognition in the wild* (Cambridge, MA: MIT Press, 1995) and Yrjö Engeström and David Middleton (eds), *Cognition and communication at work* (Cambridge: Cambridge University Press, 1996).

34 For an example of a mutually supporting intellectual network a long way from cities and institutions, see Leonie Hannan, 'Collaborative scholarship on the margins: An epistolary network', *Women's Writing*, 21:3 (2014), pp. 290–315.

35 See digitisation and mapping projects such as 'Cultures of knowledge': www.culturesofknowledge.org (accessed 26 July 2019). There are a wide range of publications that map intellectual networks through correspondence; examples include Carol Pal, *Republic of women: Rethinking the Republic of Letters in the seventeenth century* (Cambridge: Cambridge University Press, 2012); and for a more encompassing treatment of networks and collaborative knowledge-making, see Paula Findlen (ed.), *Empires of knowledge: Scientific networks in*

the early modern world (London: Routledge, 2018); Hanna Hodacs, Kenneth Nyberg and Stéphanie van Damme (eds), *Linnaeus, natural history and the circulation of knowledge* (Oxford: Voltaire Foundation, 2018).

36 Steven Shapin, *A social history of truth: Civility and science in seventeenth-century England* (London: University of Chicago Press, 1994).

37 Robert Iliffe, 'Material doubts: Hooke, artisan culture and the exchange of information in 1670s London', *British Journal for the History of Science*, 28 (1995), pp. 285–318.

38 Andrew J. Bennett, 'Expressivity: The Romantic theory of authorship' in Patricia Waugh (ed.), *Literary theory and criticism: An Oxford guide* (Oxford: Oxford University Press, 2006), pp. 48–58.

39 For an illuminating examination of the Society for the Encouragement of Arts, Manufactures and Commerce in particular, see Mat Paskins, 'Sentimental industry: The Society of Arts and the encouragement of public useful knowledge, 1754–1848' (PhD thesis, University College London, 2014).

40 Jennifer Uglow, *The lunar men: The friends who made the future* (London: Faber, 2002); see also Peter Clark, *Sociability and urbanity: Clubs and societies in the eighteenth-century city* (Leicester: Victorian Studies Centre, 1986); James Kelly and Martyn J. Powell (eds), *Clubs and societies in eighteenth-century Ireland* (Dublin: Four Courts Press, 2010); Amy Prendergast, *Literary salons across Britain and Ireland in the long eighteenth century* (London: Palgrave Macmillan, 2015).

41 See Markman Ellis, *The coffee house: A cultural history* (London: Weidenfeld and Nicolson, 2011); Berry, *Gender, society and print culture*; and Jan Golinski, *Science as public culture: Chemistry and enlightenment in Britain, 1760–1820* (Cambridge: Cambridge University Press, 1992).

42 Richard Coulton, '"The darling of the Temple-Coffee-House Club": Science, sociability and satire in early eighteenth-century London', *Journal for Eighteenth-Century Studies*, 35:1 (2012), pp. 43–65.

43 James Tierney, 'Periodicals and the trade, 1695–1780' in Michael F. Suarez and Michael L. Turner (eds), *The Cambridge history of the book in Britain*, vol. 5, 1695–1830 (Cambridge: Cambridge University Press, 2009), p. 483 (pp. 479–97).

44 For a later period, see Secord, 'Science in the pub'.

45 Secord, 'Knowledge in transit'; there has been an extensive discussion of the 'popularisation of science' in the nineteenth century, see for example: Bernard Lightman, 'Marketing knowledge for the general reader: Victorian popularizers of science', *Endeavour*, 24:3 (2000), pp. 100–6.

46 Karen Harvey, *The little Republic: Masculinity and domestic authority in eighteenth-century Britain* (Oxford: Oxford University Press, 2012).
47 For a discussion of the centrality of the American colonies to cultures of curiosity and natural history, see Susan Scott Parrish, *American curiosity: Cultures of natural history in the colonial British Atlantic world* (Chapel Hill, NC: University of North Carolina Press, 2006).
48 Elaine Leong, *Recipes and everyday knowledge: Medicine, science and the household in early modern England* (Chicago, IL: University of Chicago Press, 2018); Joan Thirsk, *Food in early modern England: Phases, fads, fashions 1500–1760* (London: Hambledon Continuum, 2006).
49 Jane Whittle, 'A critique of approaches to "domestic work": Women, work and the pre-industrial economy', *Past & Present*, 243 (2019), pp. 35–70; see also Jane Humphries and Jacob Weisdorf, 'Unreal wages? Real income and economic growth in England, 1260–1850', *The Economic Journal*, 129:623 (2019), pp. 2867–87; in another context, Francesca Bray has used a feminist reading of histories of women and technology to unsettle assumptions about domestic space and gendered power in late Imperial China, see Bray, *Technology and gender: Fabrics of power in late Imperial China* (Berkeley, CA: University of California Press, 1997).
50 Dorothy Marshall, *The English domestic servant in history* (London: Historical Association, 1949); Joseph J. Hecht, *The domestic servant class in eighteenth-century England* (London: Routledge & Kegan Paul, 1956); Bridget Hill, *Servants: English domestics in the eighteenth century* (Oxford: Clarendon, 1996).
51 Steedman, *Labours lost*, p. 14.
52 Sara Pennell, *The birth of the English kitchen, 1600–1850* (London: Bloomsbury Academic, 2016); Elaine Leong, 'Collecting knowledge for the family: Recipes, gender and practical knowledge in the early modern English household', *Centaurus*, 55:2 (2013), pp. 81–103; Craig Muldrew, *Food, energy and the creation of industriousness: Work and material culture in agrarian England, 1550–1780* (Cambridge: Cambridge University Press, 2011); Harvey, *Little Republic*.
53 For example, Bridget Hill, *Women, work and sexual politics in eighteenth-century England* (Oxford: Basil Blackwell, 1989); Isabelle Baudino, Jacques Carré and Cécile Révauger (eds), *The invisible woman: Aspects of women's work in eighteenth-century Britain* (Aldershot: Ashgate, 2005); Nicola Phillips, *Women in business, 1700–1850* (Woodbridge: Boydell and Brewer, 2006); Harvey, *Little Republic*.
54 As Jane Whittle highlights, male domestic labour is often described as 'farming' or 'construction' and thereby acknowledged as part of the

wider economy, whereas women's work at home is often categorised amorphously as 'housework' or as 'care', neither of which are considered worthy of inclusion, see Whittle, 'Critique', p. 43.

55 British Library (hereafter BL), Trumbull Papers (hereafter Trumbull), Add MS 72516: Anne Dormer to Elizabeth Trumbull, 10 September *c.* 1687.

56 BL, Trumbull, Add MS 72516: Anne Dormer to Elizabeth Trumbull, 22 June *c.* 1687.

57 See, for example, Linda Siedel's exploration of Jan van Eyck's fifteenth-century altarpiece, which emphasises that material knowledge learned at home could both connect artists with other forms of expertise and networks and find outlets in artistic practice itself, 'Visual representation as instructional text: Jan van Eyck and the Ghent altarpiece' in Smith and Schmidt, *Making knowledge*, pp. 45–67, one example being 'Alum' – a resin commonly used in the domestic treatment of illness, but also a binding agent for paints.

58 Steven Shapin, 'The house of experiment in seventeenth-century England', *Isis*, 79:3 (1988), pp. 373–404; see also Deborah E. Harkness, 'Managing an experimental household: The Dees of Mortlake and the practice of natural philosophy', *Isis*, 88:2 (1997), pp. 247–62, especially for the role of a wife in the domestic 'business' of science; and Donald L. Opitz, Staffan Bergwik and Brigitte Van Tiggelen (eds), *Domesticity in the making of modern science* (London: Palgrave Macmillan, 2016).

59 Simon Werrett, 'Recycling in early modern science', *British Journal for the History of Science*, 46:4 (2013), pp. 627–46.

60 Werrett, *Thrifty science*.

61 Opitz et al., *Domesticity*; Mary Terrall, *'Catching nature in the act': Réaumur and the practice of natural history in the eighteenth century* (Chicago, IL: University of Chicago Press, 2014), p. 26, see also pp. 44–78; see also Clare Hickman on gardens as important locations for the medical practice of physicians and spaces that illuminate connections between medicine, chemistry, botany and agriculture, 'The garden as a laboratory: The role of domestic gardens as places of scientific exploration in the long 18th century', *Post-Medieval Archaeology*, 48:1 (2014), pp. 229–47.

62 Leong, *Recipes*; see also Michelle DiMeo, 'Lady Ranelagh's book of kitchen-physick? Reattributing authorship for Wellcome Library MS 1340', *Huntington Library Quarterly*, 77:3 (2014), pp. 331–46; Lucy J. Havard, '"Preserve or perish": Food preservation practices in the early modern kitchen', *Notes and Records*, 74 (2020), pp. 5–33; see Pennell, *English kitchen* for the experimental potential of this room and also for

the interconnected nature of production, consumption, knowledge and technology.

63 Alan Warde, 'Consumption and theories of practice', *Journal of Consumer Culture*, 5:2 (2005), pp. 131–53; Warde draws on the work of Andreas Reckwitz, especially 'Toward a theory of social practices: A development in culturalist theorizing', *European Journal of Social Theory*, 5:2 (2002), pp. 243–63.

64 Warde, 'Consumption', p. 135.

65 Pierre Bourdieu, *Outline of a theory of practice*, trans. Richard Nice (Cambridge: Cambridge University Press, 1977); Michel de Certeau, *The practice of everyday life* (Berkeley, CA: University of California Press, 1988).

66 Alison Blunt and Eleanor John, 'Domestic practice in the past: Historical sources and methods', *Home Cultures*, 11:3 (2014), pp. 269–74.

67 Michael Polanyi, *The tacit dimension* (Chicago, IL: Chicago University Press, 2009 [1966]), p. 4.

68 Smith and Schmidt, *Making knowledge*, esp. p. 13; focused on an analysis of didactic literature over the period *c.* 1550–1830, see Glaisyer and Pennell, *Didactic literature*, esp. p. 7.

69 Chandra Mukerji, 'Women engineers and the culture of the Pyrenees: Indigenous knowledge and engineering in seventeenth-century France' in Smith and Schmidt, *Making knowledge*, pp. 19–44.

Part I

1

Household materials and networked space

Homes are collections of objects amassed over time, some in daily use while others sit on shelves undisturbed for years. The eighteenth century is often characterised as a period of proliferation and diversification of the material world. Household inventories – lists of objects organised room by room – first gave historians the insight they sought into the longer-term changes in home comfort over this period and the role of material acquisition in that process.[1] Eighteenth-century homes were complex spaces through which people, things, materials and knowledge circulated. Masters and servants alike exercised a wide range of technical competencies and material literacies in the activities they conducted at home – using minds and hands to achieve work of both a necessary and a more exploratory nature.

By examining the circulation of materials that provisioned the home, domestic space can be seen as connected with other domestic, commercial and artisanal spaces. Through the countless people (servants, visitors, traders) and materials (fuel, foodstuffs, linen, ash) that moved through this space, the home was integrated with other local environments but also with the sprawling networks of global trade and empire. Thinking of the home as a networked and dynamic space casts a different light on the work of the home. Far from being a discrete space set apart from the main action, the home framed people's engagements with other spheres. Moreover, the household produced varied kinds of interrelated labour. This chapter shows the connections between these different forms of domestic work and argues that they created the conditions for scientific enquiry.

The archive reveals this period as one of avid household record-keeping. Manuscript collections abound with account books, recipe books and bundles of household expenses, offering an intricate record of this ubiquitous social and economic unit in eighteenth-century life.[2] The thorough recording of incoming and outgoing goods, services and money was a visible sign of orderly and thrifty household management.[3] Countless advice manuals attest to the social weight placed on achieving a shrewd use of domestic resources, a weight not evenly felt by individuals charged with this task, but a weight nonetheless. Here the evidence of the inventory and account book is crucial to assessing the affordances and demands of home on the people who lived and worked there.

To better understand the material flows of the eighteenth-century home, a range of English, Irish and North American households are examined. Whilst there were significant continuities in the material lives of elite homes across these different locations, more modest households reveal the particularities of the regions. Moreover, by examining the material culture of households with divergent local environments and supply chains, the flexibility of domestic space to facilitate and prioritise some activities over others is revealing.

In recent decades, historians have become increasingly interested in the experience of everyday life, and a focus on the household offers compelling insight into this realm. However, accessing the sensory and the affective in the documentary record poses challenges. The analytical terrain is complicated by the way that different fields understand 'experience', with cultural historians tending to emphasise ideas and discourse whilst social historians apply themselves to the detection and articulation of day-to-day practices.[4] Historians of science have also focused on networks of practice in their engagement with questions of eighteenth-century society, but have done so with a sense of a 'future-oriented' 'social imaginary' comprised of the beliefs and expectations borne out of everyday experience.[5] Nonetheless, the surviving primary evidence speaks not only to the material realities of life lived at home, but also to the energy householders committed to carefully documenting, coordinating and appraising this facet of their existence for the present and future. This chapter and the next aim to highlight household record-keeping as an important lens on a wide range of domestic practices in this era and a genre of knowledge-making in its own right.

Home 'oeconomy'

When discussing the activity and management of the eighteenth-century home, the most useful term is 'oeconomy', which referred not only to the careful stewardship of material resources but also to the virtue of running an orderly home – a unit that was often understood as a microcosm of the nation-state.[6] This linking of the household with the polity was conceptually and symbolically powerful. Meeting domestic needs in a prudent and upstanding manner thereby had meaning that reached far beyond the confines of home. Whilst the connection between oeconomy and 'improvement' has been recognised, historians of science have stressed that oeconomical productivity was always tethered to other social and moral imperatives and did not imply the maximisation of profit at the expense of these considerations.[7] In this way, surviving household accounts reveal the remnants of an interesting network of related concerns, concerns that focused on the everyday management of material resources but which had the rather larger aspiration of generating new knowledge and national prosperity.

Household record-keeping helped men and women to manage domestic production and consumption, mitigate periods of material scarcity, rein in expenditure and generate a sense of order from an unendingly busy schedule of activity. In terms of provisioning the household, account books, lists of expenses, recipe books, diaries and letters all provide insight. Even the humble list can be considered in this light, an absolutely ubiquitous form of record-keeping – whether it was used for ingredients in cooking, furniture in an inventory or sightings of birds in the garden.[8] The title page of many an eighteenth-century book reels off a lengthy, sometimes alphabetised, list of inclusions – a style that would later fall out of fashion with printers and publishers. In its mundane and ubiquitous nature, historians have overlooked the simple but effective ordering power of the list.[9]

The precise technique and format for household accounts varied widely in this period and extant examples represent a diverse written form. Even dedicated account keepers often lacked the training to produce consistent records.[10] Whilst double-entry accounting was a highly valued skill in this era, single-entry was the more common approach at home.[11] Practices of household accounting were bolstered by the 'quantitative culture' that was growing alongside rising rates of numeracy. Older concerns of reciprocity and hospitality were gradually overtaken by numerate reckoning and precepts of debit and credit in the domestic sphere in this period.[12] Household accounting can be seen as a powerful 'mode of writing' and 'representing hours of careful labour over years and years or over a lifetime' a way in which individuals represented their domestic environment and themselves.[13]

Table 1.1 shows an extract from the Household Book of Dunham Massey Hall in Cheshire and gives a sense of domestic consumption during one week in 1743.

Dunham Massey operated a mixed economy of generating some products on-site from raw ingredients whilst buying in other items from local suppliers.[14] For example, a hogshead of small beer was purchased ready-made, but hops were also bought, presumably for use in home brewing. This was a large household in the northwest of England, surrounded by extensive parkland and inhabited by the Earls of Warrington and Stamford, and had just undergone a substantial remodelling during the 1730s. Homes of this size had considerable scope for producing necessary domestic consumables from raw materials grown or reared on-site or bought in.[15]

The household account book (1797–1832) of a more modest Anglo-Irish family, the Bakers of Ballytobin in County Kilkenny, reveals a similar approach to that of Dunham Massey. The mistress of that household, Sophia Baker (née Blunden), supervised baking, dairying, stilling and the raising of some livestock and, during autumn, had to preserve enough meat and butter to see the household through the winter. Everyday items, such as tallow candles, were made at home.[16] Baker also drew on local stores and those in the neighbouring towns of Kilkenny, Clonmel and Waterford, but door-to-door pedlars also offered opportunities to secure items such as linen, lace or ribbons. For the Bakers, who held no office in government, Ballytobin was where they spent the majority of their time. However, infrequent trips to Dublin procured more costly items such as china, silver or glass wares.[17] Whilst the domestic space afforded to home production obviously varied widely according to social status, all households engaged in some productive activity.[18] Those households of the middling to upper classes that have left

Table 1.1 Extract from Dunham Massey household book, Stamford Papers, John Rylands Library

From Saturday the 26 of February to Saturday the 5 of March 1742–3	*£*	*S*	*D*
Mrs Kinaston			
42 Pounds of Butter, 17s. 6d. Eggs 13s. 6d.	1	1	
25 Partridges, 7 Fowls		11	8
Veal, Cod, Whitings, Turbats		13	
Flounders, Shrimps, sand		3	
Grocer's Bill	1	2	6
From the Dairy			
Milk, Cheeses Turkey		11	6
Fowles		1	
Used this Week		12	6
Fowls, Partridges, Turkey			
Pounds of Soap			
Butter, kitchen. Stillhouse			
Thomas Hardey			
Malt 25 Measures	4	4	
Wheat 6 Measures	1	3	
Barley 4 M. Shulling 2 Pecks		12	4
Groom, Oats, Beans		7	8
Coachman, Oats, Beans		12	8
Draughts, Oats, Colts		8	
Cows, Oats 6 Measures		6	
Partridges, Corn, s. M & 3 Pecks		2	6
Poultry, Barley 4 measures		10	
Pigeons, Corn		7	6
Two Sheep	1	10	
Brooms		1	
Mr Walton			
Quarts of red Port		3	2
Quarts of white Port		4	
Pint of sack		1	7
Quarts of Birch Wine		2	
Quarts of March Beer		4	
Barrel of Ale tap't the 3			
Hogshead of Small Beer			
Pounds of Hops		7	
Candles 6s. from the Garden 6s.		12	

Source: John Rylands Library (JRL), Stamford Papers, 'Household consumption account book', GB 133 EGR7/1/1

their records to posterity offer the most complete insight, but the material worlds of lower-class households are, however, accessible – if scarcer in the archive.[19]

Household accounts of this kind reveal their authors' ability to quantify domestic resources and map those resources onto time. The ability to calculate offered the possibility to predict and, thereby, cater for future need. Dealing in measurements that ranged from the minute to the colossal, these records offer insight into the short-term adjustments and the longer-term reckonings that householders made. The rates at which household products were acquired and used up also varied and the careful domestic manager needed to maintain a fleet of parallel calculations in mind. The household book helped her to manage this complex and ever-changing scene.

Whilst accounts reveal the quantity and diversity of materials put to use, letters offer glimpses of attitudes to these goods. At the beginning of the eighteenth century, Lady Penelope Mordaunt (neé Warburton) would regularly write to her husband, Sir John (1649–1721).[20] She wrote from 'her house over against the back gate of St. James Palace, Westminster' when he was visiting his estates in Norfolk or Warwickshire.[21] On 28 August 1701, Mordaunt's concerns were with her inability to keep weekly expenses under three pounds, despite her cook having found a supplier who would sell beef and mutton at two-pence, halfpenny and three-pence, halfpenny a pound respectively.[22] However, the Mordaunts did not buy in all of their foodstuffs and a few days later, on 6 September, Lady Mordaunt was more worried about the market value for cheese being too low for them to consider selling their own home-made product.[23] Despite this healthy home production, Mordaunt remained concerned about runaway domestic expense, although she noted dining cheaply on offal to offset the over-spend.[24]

Lady Mordaunt's worries about household expenditure and her close eye on matters of home production and consumption fit well with the ethos of thrifty home management that was extolled in the advice manuals of this era.[25] Despite their wealth, the Mordaunts still attended to the minutiae of domestic thrift. In this case, as with Ballytobin, the mistress of the household had direct oversight of these aspects of domestic labour.

In other accounts of affluent households, it is clear that members of domestic staff took up some or all of this responsibility.

The extract from the Dunham Massey household book, detailed above, reveals that Mrs Kinaston was likely the housekeeper as the kitchen, dairy and stillroom were under her purview and purchases of soap and payments to washerwomen indicate her oversight of the laundry. Other members of staff attended to other facets of home production and consumption. Nonetheless, the performance of these roles was typically coordinated by a mistress of the household whose responsibility was to ensure prudent home oeconomy.[26]

The accounts of a Dublin townhouse, located near Kildare Street and owned by James Ware (b. *c.* 1699), reveal many more bought-in products than raw materials, as compared with a country estate like Dunham Massey.[27] For example, in 1742 Ware recorded 'A Hogshead Hampshire Beer, Carriage from Chester of an ½ Hogsh^{d} beer, 25 Barells of small beer ... Small beer from Hucksters, Ale from ye Brewer, Half barell, Ale from ye Alehouse' – clearly relying on a range of suppliers, local and otherwise, to meet household needs.[28] Whilst Ware kept a record of servants' wages, none of his categories – 'Victuals', 'Drink', 'Household goods', 'Garden', 'Things in the Cellar', 'Repairs and other small matters', 'Taxes', 'Coals', 'Candles', 'Soap and blue', 'Water and washerwoman' – had any other name against them.[29] Separate, intermittent payments for the time of a gardener, bricklayer and carpenter suggest that Ware himself oversaw everyday domestic expenditure, hiring in extra help for ad hoc jobs as they arose. The extent to which a master or mistress took a hands-on approach to domestic provisioning certainly affected the dynamics of household work and the surviving documentary evidence of that work. Nevertheless, even those who delegated the vast majority of tasks to trusted servants were still expected to maintain an overview of expenditure.

To this end, on 25 March 1748, Dunham Massey took stock. The household account book records, 'Housekeepers Stok of breading and killing: 19 Fowls, 6 Turkies, 7 Geese, 3 Ganders, 11 Fowls for killing, 15 Chickens 4 Partridges'; 'Thomas Hardeys Stock of Horses Cowes &c: 6 Stable horses, 1 Saddle Mare in fole, 3 Colts, 7 Coach horses, 7 draught horses, 1 old mare in fole, 1 old blind mare, 11 milk cows, 2 barren cows, 1 fatt cow ... 2 year old calves, 2 Bulls, 20 Wathers,[30] 15 Ewes, 8 lambs, 1 ram, 2 boars, 1 sowe, 7 young hogs'.[31] This process of tracking consumption was a common one for larger estates. Although quite different in format, a 'Memorandum of the different articles of consumption for the year[s] 1783, 1784, 1785' was kept for the Dublin townhouse and grand County Kildare estate, Castletown, owned by the Conolly family. Large totals were detailed for the annual quantities of cheese, lemons, oranges and apples; Irish crabs and lobster, veal and sweetbreads, oxen, lamb, sheep and pigs; and ale and small beer brewed. The record indicates whether the family bought items in town or country and also the amounts of beef and mutton they gave to the poor.[32] These summaries provide a sense of the sheer volume of materials, goods and animals that circulated through these large households on a regular basis. They also reveal a strong oeconomical urge to account for, and sometimes restrain, the lavish spending of wealthy households.

Most accounts of middling or elite domestic consumption offer lots of detail of products that fed, clothed and cleaned the inhabitants alongside the odd status purchase of fine china or silk upholstery. But sometimes a household account can make explicit the intellectual verve of its author, and this is true of the house book for 1796 and bundles of 'bills paid' that sit in the Petworth House Archive, West Sussex. For the subject of this book, Petworth House has an interesting story to tell on account of the figure of Elizabeth Ilive (*c.* 1770–1822). She was the mistress of George O'Brien Wyndham (1751–1837), third Earl of Egremont, living at Petworth House for about fifteen years before marrying him in 1801 and becoming Countess of Egremont.[33] Her unusual life history will be discussed more fully in Chapter 6. Here, the domestic record-keeping of Petworth House is explored for its insight into domestic enquiry.

A bundle of bills from 1798, which were paid by 'Mrs Wyndham', the title Ilive adopted during her time as the Earl's mistress, reveals regular purchases of writing and drawing materials (inkstands, a mahogany desk, a flesh-coloured crayon, two dozen pencils, half a dozen black chalk, a silver pencil case) alongside travel literature, pearl-handled spoons, bone-handled knives, a plant catalogue and four eye cups.[34] Another collection of bills paid, this time between the years of 1790 and 1800 and by both the Earl and Mrs Wyndham, lists purchases of 'Botanical Magazines', books, a thermometer and a pianoforte among other more prosaic items such as cheese, bacon and oats.[35]

A house book dated 1796 is more voluble on the scale and type of enquiry taking place at Petworth at this time. On 16 January, eighteen men were paid for between one and six days each to make a reading desk and a frame for the 'Philosopher's Room', among other tasks including mending coops and fences, making gates and hewing a post in various locations across the estate.[36] Little over a month later, Egremont instructed staff to undertake 'Making Reading Desk, framing Pictures and Maps, Making Bedsteads, Hanging Doors, Putting on locks & bolts, making & putting Wooden Bottoms to Chairs'.[37] A series of entries running from March to June mention carpentry designed to create a functioning 'Silk worm Room' – a space to cultivate silkworm colonies capable of generating raw silk.[38] Mention is made of the making of a stand for a globe, a drawing table and drawing boards for Mrs Wyndham, 'a Desk to write on and to put books in to stand in Library', a pedestal for a statue and the 'Making and Canvassing [of] Boxes for Mr Ferryman & fixing his Birds &c in the North Gallery'.[39] These lists of works completed include not only the construction of specialised furniture, but also the augmenting of existing domestic space to house artworks, maps, taxidermy birds and even a colony of silkworms. There are glimpses too of the resource lavished on the hot houses, where exotic plant specimens were likely grown. They were regularly improved with new lighting, barometers ('weather glasses') and protective cases for these instruments. Whilst Petworth's master and mistress were unusually devoted to the arts and sciences, these household accounts show that home improvement was an unending process, including the necessity of mending fences alongside the bespoke design of spaces for housing collections and undertaking investigative work.

The domestic records discussed here reveal a complex ecology not only of raw materials, finished products and human labour, but also the varied and sometimes overlapping roles and responsibilities of masters, mistresses and their domestic staff. At Dunham Massey, a grand country house, a housekeeper, a house steward, alongside a retinue of maids and groomsmen kept charge of their various domains. Meanwhile, the equally grand Mordaunts kept a much closer eye on their own domestic production and consumption. At Petworth House, a wealthy and motivated Earl and his mistress adapted their estate to accommodate varied intellectual pursuits, but this evidence of expensively complex home improvement sat cheek by jowl in their accounts with the purchase of basic provisions. Every householder, however curious, had to spare a thought for their stores of salted meat and small beer.

Rooms and their uses

By the later seventeenth century, the layout of domestic space had undergone considerable change. Homes, large and small, had shifted away from the medieval format of a central, high-ceilinged hall with smaller adjoining spaces towards the proliferation of more specialised rooms and the greater use of multiple storeys. A traditional historical narrative saw early modern householders abandon their communality, characterised by masters and servants sharing beds as well as dinner tables, in favour of increasing amounts of privacy in bedchambers, closets and back parlours. Corridors helped avoid unnecessary human traffic through rooms of a more secluded nature and, by the nineteenth century, those who could afford it might separate servants and their workaday rooms from those spaces that afforded comfort and class-specific conviviality to the master and mistress of the house. Homes also became much fuller with objects of domestic utility, comfort and decoration.[40]

The literature on changing architectural plans is useful when considering the material flows of eighteenth-century homes.[41] However, it is worth recognising that considerable mutability remained in terms of the purpose domestic space was put to. A narrow focus on designated room use can result in an overly rigid understanding of room specialisation. Change was far from uniform and great variation existed in domestic room design and use, depending on both region and class. Moreover, the wealthy and powerful were not always at the forefront of new adaptation.[42] Here, rooms and floorplans are discussed but with this flexible approach to use in mind.

Floorplans provide insight into the flow of goods and people through the house. For example, in elite homes the scullery was used for washing and cleaning dishes and cooking equipment, preparing vegetables, fish or game and, therefore, it was desirable for there to be direct communication between this room and the kitchen, alongside the yard, coal cellar, wood house and ash bin. However, owing to the heat and odour that emanated from the scullery, it typically did not connect directly with spaces that contained fresh produce, such as the larder, dairy, pantry or other food stores.[43] In more modest homes, sculleries were also often housed in lean-to structures or outhouses suggesting similar preferences and also a fire safety precaution for a space which often contained a hearth.[44]

Eighteenth-century householders and servants were attuned to the relative heat and cold of adjoining workspaces in order to ensure that produce did not spoil. In the later eighteenth century, Susanna Whatman remarked in her housekeeping book that 'Butter, radishes, or anything that spoils in a hot kitchin should be placed near the parlor door, as should the cheese, to be ready to come in.'[45] For substantial country estates, Palladian architectural design promoted a 'spinal corridor' basement plan to keep domestic offices 'below stairs' whilst facilitating production, storage and serving of food and drink.[46] New wings and blocks accommodating the productive offices of the household were often added in this period to reduce both the risk of fire and the drift of kitchen odours into smarter parts of the house.[47]

More modest households also underwent change, as N. W. Alcock's detailed study of the well-preserved Warwickshire parishes of Stoneleigh and Ashow shows. Drawing on a rich supply of probate inventories, parish records and extant architecture, Alcock demonstrates that room use altered across the social spectrum between the sixteenth and eighteenth centuries. This sample is comprised principally of yeomen, wealthy husbandmen, craftsmen and the lower gentry who had enough material wealth to leave their mark on the record. Their homes witness the predictable shift away from a hall as the heart of the home and, similarly, the proliferation of furnishings with ramifications for domestic comfort. In this sample, five-room houses usually included a kitchen and often also a pantry or buttery and even a dairy.[48] In six and seven-room dwellings, the inclusion of a dairy and a buttery or pantry became more likely and stables were also a frequent addition.[49] Farmhouses commonly encompassed a couple of service rooms (likely a dairy and a pantry) alongside the kitchen, and the larger examples also included a brewhouse and a cellar.[50] In these Warwickshire villages 'A cheese chamber was almost universal', although these rooms often stored other kinds of goods, such as wool and corn.[51]

At New House Farm in Stareton, built in 1716, the ground floor included the productive rooms of kitchen, dairy, pantry and brewhouse, with a back parlour over a cellar space. The first floor featured the commonplace cheese chamber (over the brewhouse), and a further four chambers – one listed as the 'best' with an adjoining closet.[52] Such a house prioritised the functional roles of the home, giving up half of all domestic space to the making and storing of consumable goods, a good deal of space to retirement and relatively limited house room for entertaining.

Ursula Priestley and Penelope Corfield's study of room use in Norwich based on 1,408 probate inventories and archaeological evidence offers an urban comparison with Alcock's rural village. Norwich homes experienced a growth of domestic material culture, especially chairs and tables, suggesting a general increase in comfort. The expansion in kitchen furniture also implies that this room had become a key living space for the family, displacing the traditional hall and, perhaps, reserving the parlour for entertaining guests. This study notes the presence of books in the kitchen, especially Bibles, pointing to the use of the room for family prayers and further corroborating a sense of the kitchen as a living as well as a functional space.[53]

By the eighteenth century, half of the Norwich sample had a washhouse with a hearth, in other words the likelihood of a heated copper for use for laundry. Whilst the prominent textile industry of the city might have prioritised this helpful domestic facility, it is also telling that between 1705 and 1730, 19 per cent of these washhouses appear to have been used simultaneously for brewing.[54] This evidence shows the flexible use that could be made of apparatus for heating, cleaning and processing domestic resources. This was particularly important when a home was also a business and 30 to 50 per cent of households incorporated a ground floor shop or working rooms dedicated to craft activities between 1580 and 1730. Many also used garrets as spaces for weaving, demonstrating the way homes accommodated a wide range of types of work.[55]

It is worth noting that vernacular architecture and living conditions in Ireland were different from those of both rural and urban England. Whilst the Anglo-Irish Ascendancy benefitted from large estates, country homes and townhouses on a similar scale to their British counterparts, the general population in Ireland endured much more basic housing. Before the mid-nineteenth century, most Irish homes were built by their owners out of materials ready to hand.[56] In earlier periods, timber-frame structures were common and usually covered in sods, clay, straw or wattle. After extensive deforestation in Ireland, stone and mortar constructions became dominant and, in some regions, dry-stone walling was preferred.[57]

Over the eighteenth century, Irish homes were most often single-storey, rectangular buildings of a single room in depth with a loft used for storage. As such, they had less scope for specialised spaces of home production than many households across the Irish sea. By the 1800s, 'more than half of all vernacular houses were four bays long with three windows', the doorway and one window belonging to the kitchen.[58] Usually, other rooms were bedrooms but in the larger home, there was often a parlour. Whilst vernacular Irish households were typically built on a smaller scale than their English equivalents, the kitchen was still the most important room. However, it is important to recognise the heterogeneity of Irish buildings and their responsiveness to specific environments in terms of design and use of materials.[59]

Taken together, this evidence of rural and urban, English and Irish homes acts as a caution against taking the arrangements of elite homes as the model for room use in this period. Despite the greater scope of these establishments to achieve desired ends, their design and use were often divergent from ordinary homes and a 'trickle-down' model of change does not fit the rural and urban studies discussed here.

An interesting comparison with patterns of room use present in Britain and Ireland is the homes created by migrants to the east coast of America in this period. An enlightening sample of over ten thousand inventories for properties in Chester County, Pennsylvania, exists for the years 1682–1849. This was one of the first three counties formed by William Penn under royal charter, but the majority of new householders in the 1680s came from the British Isles, including many English and Welsh Quakers and Baptists.[60] Presbyterians and Anglicans followed and by the first decades of the eighteenth century, they were joined by Irish Quakers and Ulster Presbyterians.[61]

Like the Norwich study, these people were largely tradespeople or artisans, including large numbers of weavers, blacksmiths, carpenters and masons, a good number of coopers and shoemakers, tailors and doctors, alongside the requisite tavern-keepers, shop-keepers and distillers and a handful of painters, plasters and tutors.[62] Of course, the distinct economic conditions of America's eastern seaboard underpinned consumption practices. This was a credit-dependent tobacco economy and a frontier society, where the wealthiest had the easiest access to merchandise, not only through ease of credit, but also their ability to travel and their far-reaching networks of association.[63] These were diverse migrant communities adapting to their new climate and local natural resources outside of the bounds of major urban centres. Their buildings, furniture, foods and social relations were correspondingly heterogeneous, representing an accommodation between cultures of origin and local conditions.[64]

In Chester County, most homes had several spaces that could accommodate the production of consumable goods and the storage of raw ingredients and specialised equipment. Practices of room use common in Britain and Ireland are visible here, especially in terms of longer-term change. One of the most ubiquitous rooms was the kitchen, which by the late 1600s was usually found within the main house. Some households had an additional 'back kitchen', 'wash kitchen' or 'out kitchen'. There were very few designated 'dining rooms' in this community before 1830, the kitchen performing this function.[65] The lists of objects found in ten inventories dating 1688 to 1817 consistently include equipment (such as a jack, spit or tongs) and vessels (iron pots or 'fire vessels', iron kettles) for cooking on or over the fire.[66] Practical tableware and cooking pots made from earthenware appear in seven of the ten inventories and pewter in all but two. These inventories also reveal a prevalence of items associated with an earlier period of British domestic furnishings, such as pewter (as opposed to china). Two householders still owned trenchers – flat, wooden eating surfaces reminiscent of much earlier table settings in Britain. Only Benjamin Shaneman of Vincent owned anything finer, 'six Queensware plates', and his inventory was dated at the later end of the period – 1817.[67] The kitchens were often furnished with specialised equipment including colanders, funnels and kneading or dough troughs. Henry Camm of Newtown even had a still in his 1758 kitchen.

These kitchens were spaces of food production, but they also offered tables, chairs and stools – sometimes a couple of armchairs – for families to sit, eat and warm themselves by the fire.

The presence of other materials including lumber, wool, linen yarn, wheat and flax implies a broader range of home production, consumption and construction. Many inhabitants of Chester County made good use of the cooler temperatures offered by cellars to store provisions, such as wine, beer, cider, salted meats, pickles, preserves and cheeses. In Pennsylvania's hot summers, cool storage must have been a valued household attribute.

Like rural Warwickshire, these homes did include sociable rooms like parlours or sitting rooms, but they dedicated more space to practical matters.[68] In these houses, rooms are put to multiple uses, but over the course of the 150 years covered by the inventories, increasing specificity is visible. That said, smaller domestic spaces naturally offered less scope for room specialisation in these busy, productive homes – plates and guns, beds and lumber might well jostle alongside each other for limited house room.

These sources reveal the broad range of material processes, from cheese-making to stilling, that could be comfortably accommodated by domestic space in this era, whether that household was in an English provincial city or a newly built colonial American home. Local conditions mattered and shaped the material worlds of communities separated by many thousands of miles. However, deep continuities also existed. The more complex arrangements secured by the very many lodgers and transient tenants of large urban areas – where even access to a heat source was not guaranteed – are beyond the scope of this chapter. However, it is the potential these domestic spaces held for enquiry that concerns this study. The potential lies in the details of the materials, processes and spaces outlined here.

Working rooms: kitchen, brewery and stillroom

Having considered the shifts in the type and use of common rooms in homes across the British world, it is worth taking a closer look at the material culture of key productive rooms. Of all domestic spaces, the kitchen exemplified the incredible diversity in materials and processes administered on a daily basis. Most kitchens afforded the curious householder a good heat source, a variety of specialised apparatuses and often a large space – or at least a substantial table – to work with. In 1739, the Gells of Hopton Hall in Derbyshire had a well-equipped kitchen including equipment to make the most of the roasting potential of the fireplace ('Three coal rakes', 'two racks with hooks', 'six large spitts & one bird spit'), pans that indicated the existence of a stove or hot plate ('nine Sauce pans; four stew pans; five brass pans; four fish pans; two leaden fish pans') and a wide variety of other items from wooden scales to an egg slice.[69] In A. W. Baker's household account book for Ballytobin House, 'A list of Kitchen Things' includes 'preserving Pan Copper', alongside a range of meat cutting, butchering and mincing equipment.[70] Likewise, an 1825 inventory of Styche Hall in Shropshire revealed the kitchen packed full of equipment that would facilitate the production of diverse consumables, including '11 Copper Stew pans & preserving pan', 'Two tin fish strainers', a 'Lanthern & two reflectors', a 'Cradle Spitt 20 Meat hooks in ceiling' and 'Two loafs of Sugar'.[71] This elite household also benefitted from a larder, scullery, brewhouse, malt room and salting room, each offering further apparatus and supplies for bespoke provisioning.

During this period, kitchens in larger households shifted from having an open fire to becoming a closed hearth and, later, a range. This change in format provided the cook with a smoke-free kitchen but reduced the flexibility of use of the fire itself, especially for experimental purposes. However, as the fire became enclosed, other adjacent spaces, such as the scullery, were more commonly found in house design.[72] Interestingly, publications from the earlier part of the period reflect the flexible use of the kitchen. John Rudolph Glauber's expensively printed and bound *The works of the highly experienced and famous chymist, John Rudolph Glauber: containing, great variety of choice secrets in medicine and alchymy* (1689) included notes on 'the Extrinsecal use of the Spirit of Salt in the Kitchen' alongside other guidelines for alchemical procedures.[73] Glauber recommended the use of 'spirit of salt' in place of vinegar or lemon juice as a means of rendering the flesh of an old hen 'as tender as a chicken' when boiled with spices, water and butter.[74] This book reveals a contemporary association between activities such as cooking and chemistry, the frontispiece boasting that 'the Art of Chymistry is very useful and highly serviceable in Physick, Chyrurgery [surgery], Husbandry, and Mechanick Arts' as 'long since evinced by the Excellent Mr. Boyl[e] ... in his Experimental Philosophy'.[75]

Another office of home production that required a dedicated space, specialised equipment and a skilled practitioner was the brewhouse. Brewing had been a significant domestic activity for many centuries; it was traditionally women's work, but by the eighteenth century, the brewer in large households was much more likely to be a man.[76] There were regions where female expertise in brewing endured, such as the Chesapeake in America where they made most alcoholic beverages into the late eighteenth century.[77]

Brewing was affected by the seasons, with most ale production taking place in the non-summer months in Britain and Ireland. Small beer production relied on regular brews because it did not keep as well as stronger ales, which could be stored for up to a year without spoiling.[78] Country house brewing was conducted on a large scale and at the turn of the nineteenth century, domestic brewhouses still accounted for half of all British beer production.[79] Whilst essential to the household, brewing demanded expertise in the complex field of fermentation and such skills were highly valued amongst domestic servants throughout this period.[80] Beer was also an important source of energy for the labouring class.[81] A beer allowance frequently substituted for part of a servant's wages and so the domestic production of beer remained fundamental to the economy of a large household in this period.[82]

The 1825 household inventory for Styche Hall, in Shropshire, reveals the following as contents of the 'Brewhouse':

> Brewing furnace, nearly new stack lead Curve & Grate, Iron furnace & appendages, Five Mashing Tubs, Three Large oval coolers, Six small Round Coolers, Rince Tub and Gasser, Tun[ing] dish Gaun & pail, Cleansing scieve & Mash, Rules, Old Barrel & Small Cask, Oven Peel Scraper & fork, Water Trough & Spout, Four large stillages, One Bench, New Round Tub & old ditto, Iron Water dish.[83]

A 'Brew House' detailed in a 1743 household inventory for Compton Place in East Sussex reveals an elaborate set-up, starting with a 'Large Brewing Copper' accompanied by a diverse range of vessels including mash tubs, coolers, a rinsing tub, a washing copper, troughs and a malt mill. This entry also lists two ladders, two pulleys and ropes, one rake and a mashing staff – evoking the scale of the enterprise, whereby ladders were required to reach the mouth of the large copper or 'high wash Tubs' and implements on long handles to stir and remove surface detritus from the brew.[84] The Gells of Hopton Hall possessed smaller-scale facilities, but they still included items for heating, mashing, cooling and pouring liquid from one receptacle to the next.[85] This evidence emphasises the technical needs of brewing at scale, with two kinds of furnace at the heart of the Styche Hall operation.

Bakehouses and brewhouses were sometimes built adjacent to one another so that one furnace could facilitate both activities, for example at Foremark in Derbyshire (built 1759–61). A wealthy household would have produced beer of three different kinds on a regular basis and this required adjustments in the process to achieve the desired variations in flavour and alcoholic strength.[86] As Lord Mordaunt advised his wife, Lady Penelope, there were strategies for dealing with over-production: 'Pray consider that wee do not want Ale when I come but rather Brew againe so to have some Bottled.'[87] Similarly, the Irish Quaker, author and diarist, Mary Leadbetter (1758–1826), recorded that 'Thomas Bewley and I bottled ale' on 4 October 1791.[88] Unlike other regular facets of home production, brewing offered a variety of options for short- and longer-term preservation and storage.

Another working room with specialised apparatus was the stillroom. This space was of particular importance in the production of remedies and luxuries. As the name suggests, it contained a still or alembic for distilling liquids – heated by a furnace (see Figure 1.1).[89] In an 1819 inventory, Dunham Massey Hall in Cheshire listed a 'Still House' containing '2 Tables & 2 Chairs, 48 Bottles of vinegar, Quantity of old Glass, Still, Cupboard, 2 stools & Butlers Tray.'[90] The large numbers of glass containers are in keeping with a place that produced a variety of 'distilled' products that might be used in small quantities over time. Research on eighteenth-century recipe books has revealed a range of descriptors for this facility, most commonly referred to generically as a 'still', but also commonly as a 'limbeck' or 'cold still' and more rarely as a 'glass still', 'rose still' or 'bain marie'.[91]

Having traditionally been used for extracting the potent aspects of plants to produce health-giving medicinal ingredients, by the seventeenth century stillrooms were routinely also used for making and storing confectionery. The reason these functions were combined was partly because there was overlap in the techniques of production of health-giving herbal waters and

Figure 1.1 Housekeeper in her stillroom. Courtesy of the Wellcome Collection. Public domain.

celebratory spiced cordials.[92] At this time, the stillroom was also largely the domain of the mistress of the household, which designated the higher status of stilling as compared with cooking, curing or cheese-making.

Mary Evelyn (*c.* 1635–1709), wife of the famous diarist John Evelyn and a regular at Court, remarked that she had 'the care of piggs, stilling, cakes, salves, sweet-meats, and such usfull things' in 1674.[93] Whilst it is reasonable to question the extent to which

Evelyn's engagement with all of these aspects of domestic production was hands-on, the connection drawn between stilling and the creation of salves and sweetmeats was genuine. As Evelyn argued in a letter to a friend, the priorities of a wealthy mistress were 'the care of Childrens education, observing a Husbands commands, assisting the sick releeving the poore, and being serviceable to our friends'.[94] Assisting the sick by providing homemade medicines and entertaining visiting friends with lavish banquets both required time spent in the stillroom. Medicinal recipes that relied upon distillation were often collated in recipe books with other domestic tasks dependent on similar chemical processes, rather than appearing next to other medicinal remedies. This underlines the importance of technique in the ordering of these domestic recipes.[95]

At the end of 1778, the household and personal expense accounts of Jane Creighton, First Baroness Erne of Sackville Street in Dublin, revealed a cost of £1:1:18 'For sweetmeats made at home'.[96] In seventeenth- and eighteenth-century British and Irish society, upper-class households produced such confectionery for more elaborate dinners for invited guests. Sweetmeats, 'marchpane'[97] confections and jellies would often have adorned a banqueting table. A menu created for the Gells of Hopton Hall for a dinner on 30 December 1752, for example, offered a range of deserts including 'Dry'd Sweetmeats' alongside brandied peaches, syllabubs and other fresh and candied fruits.[98] If not bought at great expense from a confectioner, these showy sweet treats were made by the mistress of Hopton Hall herself. However, whilst remedies and sweets may have emerged aplenty from the stillroom, this space offered the curious individual a wide scope for experimentation with materials and chemical processes.

Stillrooms have not attracted much scholarly attention, most likely because they fell out of use at the end of the period and have not survived the household improvements of subsequent centuries.[99] During the eighteenth century, a still became more likely to be housed in a kitchen, buttery, closet, hall, parlour or brew house.[100] In fact, there appears to be only one extant stillroom in England, at Ham House in Surrey, although spaces that originally housed a still do survive, including the example in Figures 1.2 and 1.3 at Strokestown Park, County Roscommon in Ireland.[101] There is quite a bit to untangle in these images from the Irish Architectural Archive. In the seventeenth century, this room was a reception room, with a grand plasterwork

Figure 1.2 Photograph of Strokestown Park's former stillroom, featuring the fireplace and over-mantle (1987). Courtesy of the Irish Architectural Archive.

over-mantle. However, when Strokestown Park was substantially remodelled in the 1730s, it was repurposed as a stillroom.[102] Whilst the neo-Palladian redesign included wings for the productive offices of the house, including the kitchen and stables, it is possible that the beautiful plasterwork of this former reception room marked the space out for an elevated component of home production such as stilling. Of course, the heat source itself may have recommended this room for this purpose and its presentation in these images from the 1980s reveals walls lined with cupboards, which could have been added during its conversion to accommodate essential glass vessels. So, whilst it is difficult to be definitive about the extant architectural evidence, there are interesting indications here that a stillroom had a rather higher status than other facets of home production.

Figure 1.3 Photograph of Strokestown Park's former stillroom, featuring fitted cupboards (1987). Courtesy of the Irish Architectural Archive.

Around the house

In examining the household's materials, equipment and space, the garden should not be forgotten. Obviously, large country estates had vast acres at their disposal for farming, husbandry, cultivation and leisure. However, many eighteenth-century householders had access to some outside space where animals could be kept or plants grown and these were put to the service of the kitchen. Famously, the seventeenth-century diarist John Evelyn took a keen interest in horticulture and forestry, publishing *Sylva: or, a discourse of forest trees* in 1664 and substantially augmenting later editions in 1670, 1679 and 1706.[103] In his own time, Evelyn's garden at Sayes Court in Deptford was one of the best known in England.[104] His work with plants and trees prefigured the huge growth in interest in botany over the course of the eighteenth century and the significant shifts in understanding plant life of that later period. As many historians have shown, observing and documenting the marvels of nature became a popular pursuit in the British Isles – fuelled by a print culture that disseminated intriguing news of 'exotic' foreign flora and fauna encountered through networks of trade and empire. The garden has also been identified as a space of experiment for medical men and natural philosophers alike.[105]

Careful domestic oeconomy embraced the garden as well as the kitchen and pantry, as James Ware's meticulous household accounts show. Reporting on the expenditure relating to his Dublin townhouse, Ware revealed that his urban garden accommodated a wide range of activity, from the growing of kidney beans to the overwintering of valuable fruit in the apple loft and the construction of an arbour.[106] This city garden supported a wide range of provisions for the dining table; in March 1741, Ware reported the carriage of currant trees from the country to plant in his garden, alongside the planting of other fruit trees. In the same year, asparagus roots and cauliflowers were bought to grow. Bills for seeds appeared yearly in these accounts and also entries for dung, a spade and a rake, lime and sand and a good deal of paid labour, including thirty days of a gardener's time in 1741.[107] His accounts reveal a well-resourced and active kitchen garden in the heart of a busy city.

Lady Penelope Mordaunt's careful management of her English household also extended to the garden, and her letters to her husband reveal her experiments with cultivating non-native plants. Like Ware, Mordaunt often wrote from her London residence, confirming the use of the more limited outside space adjacent to townhouses for growing fresh produce. On 26 August 1704, Mordaunt reported 'I have saved ye seeds of ye two melons, but I think nether of them good' and on another occasion, having received some melons, pears and two nectarines from the country, she noted, 'I will be shuer to safe ye Melone ceeds, but I think I can send down beter seeds for ye Melan I think is two waterish.'[108] She also intended 'if there be any figs to be had' to send her husband some dried ones 'for ye are very holsom'.[109] Her discussion of planting these seeds sat amongst a litany of details about her careful household provisioning, whether that was reporting on current stocks of coal or ensuring her husband had the domestic comforts he needed when away from home.

In domestic record-keeping, the garden and estate were sometimes treated as a sphere separate enough to deserve their own record book. For example, in the archive relating to Dunham Massey Hall, the garden accounts sit apart from the household goods, in four large, hidebound volumes of their own.[110] For other domestic record-keepers, the productive function of a kitchen garden or fields ensured their inclusion within the main household accounts. Regardless of the organisation of household accounting, the material world of the home did not stop at its threshold. It commonly incorporated a traffic of goods that extended to gardens, farmland, neighbourhood outlets and beyond.

Conclusion

This chapter has explored the diversity of commodities, furnishings, equipment, forms of labour and spatial arrangements that comprised the home in this era and considered how these varied according to class and location. On the one hand, the endless lists of goods presented in domestic accounting are revealing of the great variety of materials put to use at home. On the other, household plans and inventories provide a sense of the way space was occupied and used; letters and life-writing offer further qualitative detail of home production, use of space and – crucially – the preferences of those who undertook or oversaw household work. Whilst account books leave many of the historian's questions unanswered, in the time they were written they offered their authors a powerful tool of oeconomical control. They were the means by which people managed their everyday lives, but they were also a lasting record of, and reckoning with, the material resources of life. They obliquely recognise the imprudent overspending on luxury items or the seasonal lack of fresh fruit. Annual summaries assumed a larger meaning, delivering the cumulative effect of many, small decisions in the hefty unit of tonne, barrel or carcass.

It was the interaction of material resources that represented the key to successful provisioning. As a result, householders were intent upon the constant and ever-shifting challenge of undertaking measurements accurate enough upon which to predict need and thereby provision adequately and on a budget. These domestic records speak to a prevalent cultural concern in this period, one of categorisation, classification and control: an oeconomical urge that predicted the activity of chemists as much as it did the concerns of a housewife.[111]

Taken together, these sources indicate not only the thrifty oeconomy at work in many homes across the British world but also the incredible weight of material knowledge that was necessary for this task. These records also represent a form of material knowledge in their own right. They illuminate complex and interlocking domestic dynamics and the ways in which homes connected with other spaces and supply chains. The knowledge of home was similarly networked and relational; the story of one home's resource management was the story of many materials, places and ways of knowing. In the next chapter, the discussion turns to the technique and tacit knowledge inherent in home production.

Notes

1 Lorna Weatherill, *Consumer behaviour and material culture, 1660–1760* (London: Economic and Social Research Council, 1985); see also Michael Pearce, 'Approaches to household inventories and household furnishing, 1500–1650', *Architectural Heritage*, 26 (2015), pp. 73–86; and John E. Crowley, *The invention of comfort: Sensibilities and design in early modern Britain and early America* (Baltimore, MD: Johns Hopkins University Press, 2001).

2 Karen Harvey describes the written financial account as the kind of manuscript that 'survives in the largest numbers' from this period in 'Oeconomy and the eighteenth-century house: A cultural history of social practice', *Home Cultures*, 11:3 (2014), p. 383 (pp. 375–90).

3 Lemire, *Business of everyday life*, pp. 187–226; Harvey, *Little Republic*, pp. 72–7; Margaret Hunt, *The middling sort: Commerce, gender, and the family in England, 1680–1780* (London: University of California Press, 1996).

4 Harvey, 'Oeconomy'.

5 Lissa Roberts, 'Practicing oeconomy during the second half of the long eighteenth century: An introduction', *History and Technology*, 30 (2014), p. 135 (pp. 133–48); see also Charles Taylor, *Modern social imaginaries* (Durham, NC: Duke University Press, 2003).

6 Harvey, *Little Republic*; Roberts, 'Practicing oeconomy'; and on cooperative household labour, see Amanda E. Herbert, *Female alliances: Gender, identity, and friendship in early modern Britain* (New Haven, CT: Yale University Press, 2014), esp. chapter 3, pp. 78–116.

7 Lissa L. Roberts and S. Werrett (eds), *Compound histories: Materials, governance, and production, 1760–1840* (Leiden: Brill, 2018), pp. 6–7.

8 Lists made regular appearances in print culture as well as the manuscripts of domestic life. Cynthia Sundberg Wall, *The prose of things: Transformations of description in the eighteenth century* (Chicago, IL: Chicago University Press, 2006), esp. p. 88; see also Elizabeth Yale, 'Making lists: Social and material technologies for seventeenth-century British natural history' in Smith, Meyers and Cook, *Ways of making and knowing*, pp. 280–301.

9 Sundberg Wall, *Prose of things*, p. 88; see also Lorraine Daston, 'The empire of observation 1600–1800' in Daston and Lunbeck, *Histories of scientific observation*, p. 96 (pp. 81–113).

10 Lemire, *Business of everyday life*, pp. 195–205, esp. p. 198; as Amanda Vickery has pointed out, the more competent versions of domestic accounting were probably more likely to find themselves preserved in an archive. See Vickery, 'His and hers: Gender, consumption and household accounting in eighteenth-century England', *Past & Present*, 1: supplement 1 (2006), pp. 21–2 (pp. 12–38).

11 On book-keeping practices see Mary Poovey, *A history of the modern fact: Problems of knowledge in the sciences of wealth and society* (London: University of Chicago Press, 1998), pp. 29–91.

12 Lemire, *Business of everyday life*, esp. chapter 7, pp. 187–226; see also Hunt, *Middling sort*, p. 58.

13 Lemire, *Business of everyday life*, pp. 195, 200. This issue is also discussed by Harvey, *Little Republic*, pp. 72–7.

14 See Jon Stobart and Mark Rothery, *Consumption and the country house* (Oxford: Oxford University Press, 2016), pp. 85–8 on supplies and stores and pp. 196–228 on suppliers; see also Vickery, 'His and hers', p. 25 on the wide range of suppliers in one mistress's household accounts.

15 For comparable examples of patterns of consumption in English elite homes see Stobart and Rothery, *Consumption*, esp. chapter 3: 'Practicalities, utility, and the everydayness of consumption' and chapter 8: 'Geographies of consumption: hierarchies, localities, and shopping', pp. 83–108, 229–60; for further examples of eighteenth-century book-keeping, see Steedman, *Labours lost*, pp. 67–8, 301, 306.

16 See Monica Nevin, 'A County Kilkenny, Georgian household notebook', *The Journal of the Royal Society of Antiquaries of Ireland*, 109

(1979), p. 6 (pp. 5–19); and National Library of Ireland (hereafter NLI), 'Household account book, 1797–1832', MS 42,007.

17 Nevin, 'Georgian household notebook', p. 9.

18 An exception were those individuals living in multiple occupancy lodging houses, who might have very limited access to cooking facilities. see Amanda Vickery, *Behind closed doors: At home in Georgian England* (London: Yale University Press, 2010), p. 45; Gillian Williamson, *Lodgers, landlords, and landladies in Georgian London* (London: Bloomsbury, 2021).

19 For example, a valuable study of working-class homes is Ruth Mather's 'The home-making of the English working-class: Radical politics and domestic life in late Georgian England, *c.* 1790–1820' (PhD thesis, University of London, 2016).

20 Sir John was the Fifth Baronet Mordaunt and a politician, elected to Parliament as the MP for Warwickshire in 1698 and serving until 1715; he had estates in Norfolk and Warwickshire and spent much of his time at Walton Hall in Warwickshire.

21 Warwickshire County Record Office (hereafter WCRO), Mordaunt Family of Walton Papers (hereafter Mordaunt), CR1368/1.

22 WCRO, Mordaunt, CR1368/1/33: Penelope Mordaunt to John Mordaunt, 28 Aug. 1701.

23 WCRO, Mordaunt, CR1368/1/33: Penelope Mordaunt to John Mordaunt, 6 Sep. 1701.

24 *Ibid.*

25 For a detailed discussion of the approach to home oeconomy taken by another eighteenth-century mistress and household account keeper, see Steedman, *Labours lost*, pp. 65–104.

26 For example, Thomas Hardey oversaw the livestock and Mr Walton kept the cellar; for more on the responsibilities of wives as household managers, see Amanda Vickery, *The gentleman's daughter: Women's lives in Georgian England* (London: Yale University Press, 2003), pp. 127–60.

27 James Ware was a one-time student of Trinity College Dublin and grandson of the historian Sir James Ware (1594–1666).

28 Trinity College Dublin (hereafter TCD), 'Ware household accounts book', MS 10528: 1740–86, expenses for 1742; for more on landowners' relationships with suppliers see Stobart and Rothery, *Consumption*, pp. 196–228.

29 TCD, 'Ware household accounts book', MS 10528: 1740–86, expenses for 1741.

30 'Wether' is a term for a castrated male sheep.

31 John Rylands Library (hereafter JRL), Stamford Papers, 'Household consumption account book', GB 133 EGR7/1/2.

32 TCD, Conolly Papers, MS 3951.
33 For a detailed analysis of the Petworth House Archive in relation to Elizabeth Ilive's intellectual activities, see Alison McCann, 'A private laboratory at Petworth House, Sussex, in the late eighteenth century', *Annals of Science*, 40:6 (1983), pp. 635–55.
34 Petworth House Archive (hereafter PHA), 8060: 31 Oct. 1797–Jan. 1798; the plant catalogue was likely to be *Hortus cantabrigiensi* by James Donn, which ran for thirteen editions between 1796 and 1845.
35 PHA, 'Bills paid 1790–1800', 8065.
36 PHA, 'House book', 2236: 16 Jan. 1796.
37 PHA, 'House book', 2236: 27 Feb. 1796.
38 This was a popular activity with domestic experimenters of this period and one which was encouraged by both the Society for the Encouragement of Arts, Manufactures and Commerce in London and the Dublin Society amongst other, smaller philosophical societies. For more on this subject, see Chapters 5 and 6.
39 PHA, 'House book', 2236: entries from 5 Mar.–17 Dec. 1796.
40 See Lorna Weatherill's classic study of diaries, household accounts and probate inventories for the first half of our period, showing a growth in the consumption of goods by the middling sort, *Consumer behaviour*; for discussion of the material culture of provisioning in the Irish context see Madeline Shanahan, '"Whipt with a twig rod": Irish manuscript recipe books as sources for the study of culinary material culture, c. 1660 to 1830', *Proceedings of the Royal Irish Academy: Section C: Archaeology, Celtic Studies, History, Linguistics, Literature*, 115C (2015), p. 217 (pp. 197–218).
41 For more on material flows, see Frank Trentmann, *Empire of things: How we became a world of consumers, from the fifteenth century to the twenty-first* (London: Penguin, 2016), pp. 175–90; and Chris Otter, 'Locating matter: The place of materiality in urban history' in Tony Bennett and Patrick Joyce (eds), *Material powers: Cultural studies, history and the material turn* (London: Routledge, 2010), pp. 38–59.
42 For example, Craig Muldrew has identified the switch from hall to kitchen for cooking in the homes of rural, English labourers by 1650, well before some of their elite counterparts, *Food, energy*, pp. 179–80; this is also discussed in Pennell, *English kitchen*, p. 42; likewise Ursula Priestley and Penelope Corfield note that by 1705–30 only 10 per cent of households in their Norwich probate inventory sample had halls with hearths, suggesting this room's marginalisation as a key cooking, eating and socialising space in favour of other rooms, such as the kitchen and parlour, 'Rooms and room use in Norwich housing, 1580–1730', *Post-Medieval Archaeology*, 16 (1982), p. 104 (pp. 93–123).

43 Peter Brears, 'The ideal kitchen in 1864' in Pamela A. Sambrook and Peter Brears (eds), *The country house kitchen, 1650–1900* (Stroud: The History Press, 2010), p. 15 (pp. 11–29).
44 Priestley and Corfield, 'Room use in Norwich', p. 110.
45 Christina Hardyment (ed.), *The housekeeping book of Susanna Whatman* (London: The National Trust, 1992), p. 45.
46 Peter Brears, 'Behind the green baize door' in Sambrook and Brears, *Country house kitchen*, pp. 40–5 (pp. 30–76).
47 Julie Day, 'Elite women's household management: Yorkshire 1680–1810' (PhD thesis, University of Leeds, 2007), p. 225; the properties studied are: Temple Newsam, Nostell Priory, Harewood House and Hovingham Hall.
48 Sample of sixteen five-room homes, 1701–56: twelve had a kitchen, seven had a dairy, six had a buttery or a pantry and there was one example of the following rooms: brewhouse, mill and well house. See N. W. Alcock, *People at home: Living in a Warwickshire village, 1500–1800* (Chichester: Phillimore, 1993), p. 119; a 'buttery' was the preferred term in earlier centuries and gradually gave way to the 'pantry'.
49 Sample of twelve six and seven-room homes, 1701–56 (six of each): seven had a kitchen, seven had a dairy, nine had a buttery or a pantry, two had a brewhouse and five homes included a stable. See Alcock, *People at home*, pp. 119–20.
50 Alcock, *People at home*, p. 146.
51 *Ibid.*; Warwickshire cheese was once a much celebrated local product, sold as far away as London, p. 10.
52 Alcock, *People at home*, pp. 156–7.
53 Priestley and Corfield, 'Room use in Norwich', pp. 106–7.
54 *Ibid.*, pp. 112–14.
55 *Ibid.*, pp. 109–10, 116–19.
56 Barry O'Reilly, 'Hearth and home: The vernacular house in Ireland from *c.* 1800', *Proceedings of the Royal Irish Academy: Archaeology, Culture, History, Literature*, 111C (2011), pp. 193–215.
57 Kevin Danaher, *Ireland's traditional houses* (Dublin: Bord Fáilte, 1993), pp. 21–2.
58 *Ibid.*, p. 197.
59 Deirdre McMenamin and Dougal Sheridan, 'Interpreting vernacular space in Ireland: A new sensibility', *Landscape Research*, 44:7 (2019), pp. 787–803; it is worth noting that discussion of traditional Irish houses is often steeped in nostalgia for a lost past, which can obscure important facets of these buildings' design and use and the ecological and cultural knowledges they represent.
60 Margaret B. Schiffer, *Chester County, Pennsylvania inventories, 1684–1850* (Exton, PA: Schiffer Publishing, Ltd., 1974), pp. 3–4.

61 *Ibid.*, p. 4.
62 Weavers: 210; blacksmiths: 116; carpenters: 104; masons: ninety-four; coopers and shoemakers: sixty-eight; tailors: fifty-eight; doctors: nineteen; tavern-keepers: sixty-six; shop-keepers: sixty; distillers: forty-two; painters: two; plasterers: two; tutors: two; for full details see Schiffer, *Chester County*, p. 5 (figures for 1796). On the importance of crafts in the history of America see Glenn Adamson, *Craft: An American history* (London: Bloomsbury Publishing, 2021).
63 Ann Smart Martin, *Buying into the world of goods: Early consumers in backcountry Virginia* (Baltimore, MD: Johns Hopkins Press, 2008), esp. chapter 2: 'Getting the goods: Local acquisition in a tobacco economy', pp. 42–66.
64 Smart Martin, *Buying*, p. 95.
65 Exceptions included the homes of David Lloyd (1731 inventory) and Glace Lloyd (1760 inventory), both of Chester, John Hurford's house in New Garden (1774 inventory) and the innkeeper Valentine Weaver's home in Chester (1774 inventory); see Schiffer, *Chester County*.
66 The ten inventories have the following dates: 1688, 1705–6, 1706–7, 1740–41, 1748, 1758, 1773, 1789, 1814, 1817. All of them included items for cooking on a fire, tongs and iron pots or vessels being especially common.
67 'Queensware' refers to Wedgwood creamware, an innovation aimed to imitate the desirable qualities of Chinese porcelain – it became known as 'Queensware' after Queen Charlotte ordered a service.
68 Elizabeth Collins Cromley, *The food axis: Cooking, eating, and the architecture of American houses* (Charlottesville, VA: University of Virginia Press, 2010), which argues that room use was driven by practicality and not by polite sociability; see also Sara Pennell's discussion, *English kitchen*, pp. 37–40; Lena Cowen Orlin's study of a Tudor woman of middling social status also positions the productivity of home as a spur to greater specialisation in room-use in this earlier period; see *Locating privacy in Tudor London* (Oxford: Oxford University Press, 2007).
69 Carol Barstow, *In Grandmother Gell's kitchen: A selection of recipes used in the eighteenth century* (Nottingham: Nottingham County Council, 2009), pp. 4–5.
70 NLI, MS 42,007, fos 56, 57.
71 Shropshire Archives (hereafter SA), 'Styche Hall inventory', 552/12/153: 1825.
72 Pennell, *English kitchen*, p. 44.
73 John Rudolph Glauber, *The works of the highly experienced and famous chymist, John Rudolph Glauber: Containing, great variety of choice secrets in medicine and alchymy* (London: 1689), p. 10.

74 *Ibid.*
75 *Ibid.*, n.p.
76 Christina Hardyment, *Home comfort: A history of domestic arrangements* (London: Viking Penguin in association with the National Trust, 1992), p. 82; see also Peter Mathias, *The brewing industry in England, 1700–1830* (Cambridge: Cambridge University Press, 1959).
77 Sarah Hand Meacham, *Every home a distillery: Alcohol, gender, and technology in the colonial Chesapeake* (Baltimore, MD: Johns Hopkins Press, 2009), pp. 24–5; well into the eighteenth century, the Chesapeake exhibited the brewing practices of an earlier period – maintaining the production of unhopped ale and naturally fermenting ciders whilst, on the other side of the Atlantic, the use of hops in beer and ale and the development of sophisticated stills for making stronger liquors became commonplace.
78 Sambrook and Brears, *Country house kitchen*; see also Peter Mathias, 'Agriculture and the brewing and distilling industries in the eighteenth century', *The Economic History Review*, 5:2 (1952), p. 249 (pp. 249–57).
79 Rachel Conroy, 'Country house brewing': www.pressreader.com/, 13 March 2018 (accessed 25 February 2022).
80 Sambrook and Brears, *Country house kitchen*, p. 251.
81 Muldrew, *Food, energy*, p. 66; also see information on levels of consumption and different alcoholic strengths on pp. 70, 73–83.
82 *Ibid.*, p. 253.
83 SA, 'Styche Hall inventory', 552/12/153: 1825.
84 East Sussex Record Office, SAS/CP 293: 10 Nov. 1743 – a detailed inventory of the household goods of Spencer Compton, Earl of Wilmington (*c.* 1674–1743). Spencer Compton bought East Borne estate (or Eastbourne Place) in 1724, renamed it Compton Place and rebuilt the property over the period 1726–31.
85 Barstow, *Grandmother Gell's kitchen*, p. 5.
86 Sambrook and Brears, *Country house kitchen*, p. 239; Pamela Sambrook has estimated, based on accounts dating from 1819, that the inhabitants of Shrugborough Hall consumed twenty-four gallons of beer a day; see Hardyment, *Home comfort*, p. 85.
87 WCRO, CR1368/vol. 1/20: John Mordaunt to Penelope Mordaunt, 29 Sep. 1702.
88 Riana McLoughlin, '"The sober duties of life": The domestic and religious lives of six Quaker women in Ireland and England, 1780–1820' (MA dissertation, University College Galway, 1993), p. 72.
89 See C. Anne Wilson, *Water for life: A history of wine, distilling and spirits, 500 BC to AD 2000* (Totnes: Prospect, 2006).
90 JRL, Stamford Papers, 'Household inventory', GB 133 EGR7/17/3: 1819.

91 Allen's sample consists of distillation equipment cited in twenty-seven manuscripts, which collectively contain 5,013 recipes, and the breakdown of terms is as follows: 'Still' (54 per cent); 'Limbeck' (19 per cent); 'Cold Still' (17 per cent); 'Rose Still' (5 per cent); 'Glass Still' (3 per cent); 'Bain Marie' (2 per cent); see 'Hobby and craft', p. 105.

92 C. Anne Wilson, 'Stillhouses and stillrooms' in Sambrook and Brears, *Country house kitchen*, pp. 129, 136 (pp. 129–43); diet and health were closely related concepts in the eighteenth century.

93 BL, Evelyn Papers, Add MS 78539: Mary Evelyn to Ralph Bohun, 23 Nov. 1674.

94 BL, Evelyn Papers, Add MS 78539: Mary Evelyn to Ralph Bohun, 4 Jan. 1674; see also Alisha Rankin, *Panaceia's daughters: Noblewomen and healers in early modern Germany* (Chicago, IL: University of Chicago Press, 2013); and Meredith K. Ray, *Daughters of alchemy: Women and scientific culture in early modern Italy* (Cambridge, MA: Harvard University Press, 2015).

95 Allen, 'Hobby and craft'.

96 NLI, MS 2178 – the accounts of household and personal expenses of Jane Creighton, First Baroness Erne, 1776–1799.

97 Meaning a product similar to marzipan.

98 Barstow, *Grandmother Gell's kitchen*, p. 13; pp. 11–13.

99 Katherine Allen argues that stilling continued in elite English households throughout the eighteenth century, but that there was a move away from large-scale charitable giving of medicines to local people – common in the seventeenth century, and towards smaller-scale provision for the elite household itself – giving the stillroom less prominence as a domestic office. See Allen, 'Hobby and craft', pp. 91–3.

100 Allen, 'Hobby and craft', p. 106; Anne Stobart, 'The making of domestic medicine: Gender, self-help and therapeutic determination in household healthcare in south-west England in the late seventeenth century' (PhD thesis, Middlesex University, 2008), p. 182.

101 Wilson, 'Stillhouses and stillrooms', p. 129; The Irish Architectural Archive, 'Strokestown Park, Basement Still Room', 2/94 CS4: William Garner, 1987.

102 In the 1730s, Thomas Mahon commissioned architect Richard Castle to enlarge and modify an existing house on this site; see https://theirishaesthete.com/tag/strokestown-park/; see also: www.buildingsofireland.ie/building-of-the-month/strokestown-park-house-cloonradoon-td-strokestown-county-roscommon/ (accessed 27 February 2022).

103 Frances Harris and Michael Hunter (eds), *John Evelyn and his milieu* (London: The British Library, 2003), p. 2; it was only in 1818 that John Evelyn's diary was published and, at that point, his reputation changed.

104 Harris and Hunter, *Evelyn and his milieu*, p. 11.
105 See Clare Hickman, *The doctor's garden: Medicine, science, and horticulture in Britain* (New Haven, CT: Yale University Press, 2022) and 'Garden as a laboratory'; Paula Findlen, 'Sites of anatomy, botany, and natural history' in Park and Daston, *Cambridge history of science*, vol. 3, pp. 272–89; Jan Golinski, *British weather and the climate of enlightenment* (Chicago, IL: University of Chicago Press, 2007).
106 TCD, 'Ware household accounts book, 1740–86', MS 10528: Mar. 1741; TCD, 'Ware household accounts book, 1740–86', MS 10528: Sep. 1742 – 'In Sepr the wall of the Apple loft gave way. I pull'd It down, rebuilt It, slated the little house'; TCD, 'Ware household accounts book, 1740–86', MS 10528: 1742.
107 TCD, 'Ware household accounts book, 1740–86', MS 10528.
108 WCRO, Mordaunt, CR1368/vol. 1/43: Penelope Mordaunt to John Mordaunt, 26 August 1704 and CR1368/vol. 1/48: same to same, n.d.
109 WCRO, Mordaunt, CR1368/vol. 1/48: Penelope Mordaunt to John Mordaunt, n.d.
110 JRL, Stamford Papers, 'Garden account books, 1778–1822', GB 133 EGR7/7/1–4.
111 Simon Werrett, 'Household oeconomy and chemical inquiry' in Roberts and Werrett, *Compound histories*, pp. 35–56.

2

Tacit knowledge and keeping a record

At its heart, scientific practice drives to grasp the elusive and make it tangible. This process often involves the codification of material and embodied ways of knowing. To understand the diverse engagements a wide variety of eighteenth-century people had with the world around them, the central role of experience must be acknowledged. Experience was, of course, central to eighteenth-century methods of scientific research, such as observation and experiment. The processes by which experience was conditioned and extended, the ways it worked through practices and how it was ultimately converted into evidence require examination. The aim here is to move between the mess of bodies, materials and processes and the clean and conclusive words on the page. What follows explores examples of techniques and tacit knowledge commonly employed by those who engaged in the productive work of the home. In doing so, the chapter elaborates on a range of specific domestic practices that offered householders the skills and ways of knowing that could facilitate scientific enquiry.

One of the reasons for the persistence of the false binary of hand and mind in understandings of intellectual work has to be the challenge of putting some manual processes into words. The uncomfortable relationship between forms of knowing that are based on physical manipulation and those that are based on wordy reasoning is strongly reflected in histories of intellectual life. Ideas that are primarily expressed in words, understandably, commonly find themselves in books whereas those that do not lack that 'body of work' to explain them. As art historian T. J. Clark has eloquently articulated, 'Writing automatically aims, or pretends, to be attentive. It likes details ... False vividness gives way abruptly to clever summing up.'[1] It follows that the writing down of some ideas might even distort them, make them something else, something more certain.

The household was a place of work for domestic servants, undermining the twentieth-century associations of home with cosy retreat and personal privacy. The home was often the primary place of work for masters and mistresses too, especially when a shop or workshop formed one part of the site. The record-keeping discussed in this chapter builds on the preceding one by considering a wider variety of domestic writing, including letters, recipe books and journals. What follows documents examples of the tacit knowledge necessary to run a home and investigates the ways that individuals accessed, exchanged and communicated these practice-based forms of knowledge. It also tracks the role of record-keeping as a domestic practice in these spheres of knowledge-making.

The notion of tacit, as opposed to explicit, knowledge was developed by the philosopher Michael Polanyi (1891–1976), who argued that knowing is an art and that any art is learned by practice.[2] For Polanyi, learning involved both doing and being shown how to do. Moreover, learning and knowing relied upon the social and cultural dynamics of an individual's environment and a belief in a particular phenomenon usually prefiguring any understanding of its workings. For Polanyi, all knowledge was personal and required personal participation to materialise.

With Polanyi's theory of knowing in mind, histories of science have dwelt substantially on the embodied, material and culturally embedded aspects of early modern knowledge.[3] Since the 1920s and 30s, the role of the artisan in the making of 'scientific' knowledge has been contemplated and debated and this question enjoyed renewed scholarly interrogation at the close of the twentieth century.[4] However, there remain varied interpretations of knowledge learned by doing.[5] For the purposes of what follows, knowledge learned at home was invariably of the kind that grew from practice, experiment and repetition, through learning at an expert's side, and responded to both the needs and affordances of domestic space and material culture. These were characteristics that were shared by scientific enquiry, which also often relied upon personal experience and repetition, typically in the form of observation and experiment.

Many forms of tacit knowledge did not make their way into text on the page. Nevertheless, practices of record-keeping were ubiquitous and quotidian in this period, both in the home and elsewhere. In some of these surviving texts, aspects of technique and tacit knowledge are visible. Sometimes silences in these records indicate the understanding that writers assumed readers had. All too often, the recipes that appear in domestic collections gloss over the intricacies of the process, assuming a range of competencies that are alien to the twenty-first-century reader. This chapter examines two kinds of record-keeping – domestic manuscripts concerned with provisioning and examples that document the natural world. Whilst each category is itself heterogeneous, important continuities exist across these forms of keeping a note.

Managing or conducting the many and varied tasks of home oeconomy demanded a wide range of skills and specialised knowledge. Historians have begun to identify the home as a site of experimental practice.[6] In applying the language of experiment to domestic processes, it is also worth noting Elaine Leong and Alisha Rankin's characterisation of 'experiment' as being comprised of many strains of experimental thinking rather than being one definitive practice.[7] In looking at domestic practice and knowledge in this way, two issues emerge. First, it has been difficult for historians to detect much of the unwritten tacit knowledge of home and, therefore, to fully value it in histories of knowledge.[8] Moreover, the large-scale societal change brought by processes of innovation and industrialisation in this era were steeped in the skills, techniques and tacit knowledge learned by doing. This is, of course, a well-established observation.[9] However, by bringing the details of those elusive but significant features of tacit knowledge to the typed page, a broader spectrum of knowing and a more diverse cast of intellectual actors become visible.

'Her doctrine and practice': bread-making in an elite Irish household

One vital domestic commodity that demanded especially careful treatment was the starter or 'barm' used in the leavening of bread. This substance consisted of flour, water, bacteria and yeast and often involved the transfer of material between brewing and baking – two important facets of home production in this period. Barm was a much-discussed product in eighteenth-century Ireland. The Dublin Society recorded the subject many times over in their *Transactions* and a Mrs Baker of County Kilkenny noted taking the Society's advice on keeping a large stock of barm in her 1810 household book, revealing the connections between domestic record-keeping and national debate.[10] The following example reveals the ways tacit knowledge about complex material processes was shared in person and, with difficulty, in writing.

The letter-writer was the Church of Ireland Bishop Edward Synge (1691–1762), and what follows is drawn from a large collection of published letters addressed to his daughter, Alicia Synge (1726–1807) penned between 1746 and 1752.[11] The collection is dense with detail on matters of the household and estate and Synge was clearly proud of his home production, commenting in July 1751 on his 'Fresh pleasure from the improv'd state of my Corn; which led me on to imagine it reap'd, safe in the haggard, and bread and drink from it in abundance.'[12] Synge had overseen the building of the episcopal palace at Elphin, County Roscommon, taking particular interest in planting the gardens and grounds with imported plants and trees.[13] His zeal for matters of domestic production saw him rely on the expertise of his domestic servant, but he also cast doubt on the reliability, quality and depth of her knowledge.

On 16 July 1751, Synge opened a letter with a detailed re-telling of the process of creating, maintaining and using barm. That morning, he had 'a conference ... with Jane about Bread &c.' in which she emphasised that 'the main thing is the Barm'.[14] Synge was interested in his servant Jane's technique because, in his own words, 'her Bread is Excellent, and almost constantly so. Her worst is better than the best We had last Winter.'[15] Synge was keen for bread of this high quality to be produced by the staff at his Dublin townhouse, where his daughter Alicia was resident, and hoped that she might oversee this project. However, the inherent difficulty in describing in words, rather than showing in person, was immediately apparent. Only a few lines in and Synge broke off: 'For fear of writing wrong or imperfectly I stopp'd here, and sent for Jane. My caution was not amiss.'[16] Synge had the steps in the wrong order, realising after consulting with Jane for the second time that 'the straining must be, when the Barm and Water are first mix'd'.[17]

The first revelation of Jane's practice was that the 'Best Barm is that which works out of the Vessels of Ale when drink is tunn'd [stored], the first twenty four hours.' Jane would use no other kind 'when she can help it'.[18] Once acquired, Jane kept the barm 'in a Vessel by it self unmix'd with any thing' and took a quantity from that pot on a daily basis.[19] The extracted barm was mixed with cold water, stirred and left to 'pitch', meaning settle, thus 'All dirt, and dross ... falls to the bottom, from whence she pours it off clean into another Vessel.'[20] Synge's retelling of Jane's method noted that barm taken from 'Ale or Small Beer, she holds equally good' and that this substance would keep a 'good Week or ten days'.[21] In this way, bread-making made use of the staple brews for the household, revealing the transfer of materials and material knowledge from one office of home production to another.

Jane's guidance also highlights the temporal connection between brewing and baking, the schedule of brewing providing material that could last up to ten days for the purposes of baking, before a fresh quantity would be needed.[22] As discussed below, some home bakers developed strategies to preserve their barm, making them less dependent on the brewing schedule. However, the Synge letters indicate the complex overlapping timeframes for domestic tasks. Householders and servants charged with producing everyday consumables had to bear in mind the time of gathering, accruing, preparing, making, finishing, preserving and the necessary punctuation of waiting for organic processes to do their work. Waiting time was, no doubt, swiftly reallocated to a shift with another form of production, eyes and hands regularly moving from one process to the next to make the most of each day's potential.

Preparing and maintaining the barm was an iterative process ('What she uses one day, she prepares constantly the day before') and responsive to the changing needs of the household ('Her quantity is in proportion to the Bread intended').[23] Some effort was applied to ridding the barm of detritus associated with its previous life in the brew tun: 'she ... strains immediately to get clear of Hop-Seeds &c. then she lets it pitch [settle] for a quarter of an hour or thereabouts, not longer'.[24] This process of allowing the liquid to settle was partly responsible for the clarity of the barm, but the method also included the pouring of 'blended liquor ... very carefully off into another Clean Vessel, so as to leave all dross behind'. The new vessel of barm would then be left overnight, resulting in 'the clean Barm [settling] at the bottom, from which she pours the Water off. With this thus purify'd, she makes her Bread.'[25] Having been shown the practice in person from beginning to end, Synge returned again and again to specific aspects, offering a more comprehensive description of the qualities of the materials involved, the signs of success and the elements that required personal judgement.

Tips and pitfalls in the handling of barm are identified in the next tranche of the narrative. Synge warned, 'When she pours off the liquor first from the dross, it looks as if there were little or no Barm in it.' However, this impression is misleading, and Jane reassured her audience that 'a quart of foul usually gives the next day a pint purify'd, and subsided to the Bottom from the Water'. Besides, exact quantities were not required: 'A little more or less makes no difference in the Bread.'[26] Some attention is also given to Jane's equipment, such as her use of 'Glaz'd Pans and Crocks' that 'she had for her dairy' and a vessel with 'a shallow pan, of size proportion's [*sic*] to the Liquor' into which she strained her barm and water such as 'the Dirt may pitch'.[27] Interestingly, in a subsequent letter dated 23 July, Synge requested, 'You should send back Mrs Heap's [the cook's] Vessel in which the late cargo of flow'r went up. Such things are Scarce here.'[28] Despite the episcopal residence in Elphin offering a substantial three-storey central building with the addition of two-storey wings, individual vessels were still valued and 'scarce' enough to be requested back, having travelled from country seat to townhouse. Jane's decision-making was necessarily responsive to a wide range of factors and, as such, having the right vessel to hand, was doubtless important to the smooth-running of her routine of home production.

Edward Synge's retelling of his domestic servant's practice expresses the levels of material literacy Jane required to make appropriate judgements at the many junctures in this cyclical process. For example, her description discussed how much time was 'enough for the dross and dirt to sink to the bottom, while the clean Barm continues in a floating state'. The description here is aimed specifically at enabling Synge's replication of the process. Confidence was needed in the next moment, as leaving the mixture any 'Longer time would occasion it's [*sic*] falling again to the bottom, and mixing with the dirt'.[29] Moreover, the barm harvested from the brewing process was not a uniform product. After 'another conference with her on the Subject', Synge reported Jane's response: 'Indeed, My Lord, says she, I get Barm sometimes as red as a Fox, sometimes black, full of Hop-leaves, Bog-bane, Wormwood, Artichoak leaves, and a long &c. of other like ingredients. By straining I get rid of all these.'[30] This list of ingredients is fascinating; whereas hop leaves and bog-bane were ingredients commonly used in brewing beer and ale, the others are less obvious candidates. This comment suggests that a much broader range of liquors were home-produced at Elphin or, otherwise, that Jane sometimes relied upon other kinds of fermentation to produce the barm she needed for baking.[31]

According to Synge, Jane described her use of barm as a 'doctrine and practice', highlighting both her belief in her own methods and their refinement through repetition.[32] Throughout the retelling of Jane's method, the profoundly unequal power relationship between Synge and his servant emerges, as Synge finds himself both reliant on her expertise and sceptical of her intellectual capacity to really 'know' of what she speaks. Typically, comparison and description are Jane and Synge's allies in conveying the tacit knowledge of her experience. The 'purify'd' product with which bread can finally be made is repeatedly referred to as 'white as Starch'.[33] In the second of the two letters, the comparison with starch is taken a step further, revealing Jane's sense that not only the colour compared, but also the material properties: 'barm as white and as tough as Starch'.[34] Clearly, a familiarity with the characteristics of other common domestic products and ingredients is invoked by this description. Rather than describing the barm in abstract terms, Jane relies on her audience's own material literacy as a prerequisite for carrying out the practice she describes.

Jane's own words make occasional intrusions in Synge's narrative and her pleasure at being credited for her knowledge is noted: 'I have made her very happy already, by giving her thanks from you and Mrs J for her instructions about Barm.'[35] A glimpse of a diverse palette of household ingredients emerges from the descriptions of purifying and clarifying barm: bog-bane and artichoke leaves and barm that is white as starch or red as a fox.[36] Synge notes where Jane's vocabulary departs from his own: 'Sheering, so she calls pouring', but it is hard to say whether 'the great dross, which remains in the bottom' was described as 'red like brick-dust, or darker' by Synge or by Jane herself.[37] Synge, this time using his own words, describes Jane's judgement honed by experience: 'In pouring off the first, As soon as she sees any dross rise, she stops, and leaves what remains to settle more, then pours again.'[38] He is forced to account for the responsive quality of Jane's approach, commenting, 'Sometimes she puts more Water to the dross when she thinks any good Barm is among it, and stirs agen, and after a quarter of an hour pours agen. I suppose this is when Barm is scarce.'[39] Only experience can tell Jane that there is 'good Barm ... among it'; this is her tacit knowledge and Synge is left to wonder if additional stirring is the adjusted response to a lack of barm in the pot. Distressingly for the novice, there is 'No niceity as to Quantity', in other words, no precise amount is given, and whilst Jane 'likes a good deal', Synge must discover for himself what that quantity might be.[40]

The role of knowledge in the development of Jane's technique is something that Synge certainly recognised. Nevertheless, he cast doubt on the quality and basis of that knowledge: 'To this [the whiteness and toughness of the barm] she chiefly ascribes the goodness of her Bread. How that may be I know not.'[41] Synge reported Jane as saying, 'With this, My Lord, I make all your bread, and Many a hard shift I make to get it', but criticism of her speaking too much immediately followed: 'Thus she run on, till I was tir'd.'[42] Whilst Synge took great care to make a written record of Jane's method, ultimately, he neither enjoyed hearing her speak nor entirely trusted her words, commenting, 'Either [what] she says [is] true, or the goodness of Bread depends less on Barm than We imagine.' So, despite the fact that Synge admitted 'better [bread] never was, than I have almost constantly', he was not certain that the reason for this quality was as Jane described.[43] The predictable power dynamics of class and gender operated just as firmly in this realm of knowledge-making as any other.

Ultimately, Synge urged his daughter to 'Continue therefore your Experiments, till you unravel this great mistery.'[44] This was not an idle suggestion, Synge intended to 'send you by John's Mule a Couple of Barrels of Wheat, which John is to sell. With the money you may buy more flowr for more trials. A little at a time will be best, tho' you pay more for it. A bag will be too much.'[45] Thus, for Edward

Synge, his servant's ability to create delicious bread consistently was firmly believed, through personal experience of the results, but her knowledge of the reasons for this success was doubted. Seeing for oneself, through experimentation, was considered the only reliable route to better understanding, personal experience counted.

Next, the discussion moves to consider a broader range of examples of this kind of domestic practice. Synge's detailed and extended communication with his daughter about the acquisition of tacit knowledge is an unusual archival survival. However, regular glimpses of these modes of tacit knowing, sharing of techniques and use of experiments can be identified in other household records.

Record-keeping

Domestic record-keeping generated a fine-grain understanding that afforded householders a greater level of control over their own domestic environment and finances. It is also a complicated source material, varied in its presentation and purpose, often misleading in its promise of unfiltered documentation of past actions and accumulations and simultaneously alluring as a genre that might – obliquely – speak of the self.[46] Household documenting offered the record-keeper the space to perform certain aspects of their role in terms of gender and class. Male authority was constructed through everyday domestic tasks and, especially, through the documentation of those activities. In this way, astute record-keeping was one method of maintaining patriarchal authority and enabled men to fulfil their roles as household managers and keepers of accounts.[47] Many wives also took control of domestic accounting and whilst female authority was constructed differently, many women gained power and agency through this process. The sequence of daily tasks of production, consumption and documentation shaped people's lives, demanding their attention in particular forms, trained on individual tasks and at specific moments in time, week by week, year by year. Fragments of these rhythms are captured in domestic record-keeping and offer insight not only into power relationships and the strictures of oeconomy, but also into eighteenth-century knowing and doing. The simplest forms of keeping a note can reveal the organisation not only of things and money, but also of ideas.

Home provisioning

Recipe books of this period are increasingly recognised for the evidence they contain of experimental knowledge alongside the glimpses they provide of social networks and relationships.[48] One such recipe book is that of Mary Farewell, described on the inside cover as 'her book' and begun in 1721.[49] Whilst recipe books are a diverse genre of writing, there are common characteristics and Farewell's conforms to many of these. For one, this parchment-bound book presents a mixture of culinary and medical recipes, added in no discernible order, suggesting that the book's information was accumulated over time. Farewell recorded directions 'to make Cheese Cakes', 'How to make a rice Poden', 'To make sauce for Greene Geese', 'To make a Orange Gelly pudding' and instructions for making 'Elderflower Wine' and 'Ginger Wine'. Amidst these items for the dinner table, she included a remedy 'for a weakness' and 'A Receipt to Cure the Biting of a Mad Dog'.[50] Names are added to record the origin of the recipe, a 'Mrs Phill Balgury' appears next to both 'An Orange Puding' and instructions on how 'To Dry Cherryes'.[51] Delicacies for special occasions sit alongside everyday staples and treatments for the sick or injured. Recipe books of this era record not only the practical knowledge acquired and preserved by their owners, but also the predominantly female social networks that shared and corroborated that knowledge.[52]

Another smaller, unattributed book is enclosed with Farewell's in the Derbyshire Record Office collection. It is inscribed with the impersonal title 'Receipts', meaning recipes, and comprised of lined paper folded in half, so that the lines run vertically. These features suggest a makeshift construction using materials available at hand. Nevertheless, the small pages are full of information, listing savoury and sweet dishes, condiments and remedies for a wide range of disorders for both humans and livestock.[53] The author noted many of the names of those who had offered her the recipes, sometimes in the title of the recipe itself as with Doctor Cook. However, the harvesting of recipes drew on multiple sources beyond this mistress's personal social circle. For example, 'Dutchess of Devonshire Rect for Tea Cakes' was unlikely to speak to a personal relationship with the aristocrat and more likely a recipe in popular circulation. Elsewhere, the *Derby Mercury* newspaper of 11 May 1786 is credited with a recipe to help tackle consumption, revealing access to local periodicals and a resourceful approach to gathering borrowed knowledge.[54]

Whilst the sourcing of such inclusions offers insight into networks of knowledge exchange, this chapter is concerned with what people did with this information. The role of trial and error and the need to apply well-honed judgement are obvious in the text. For example, the book admits that a method for alleviating smallpox was arrived at by accident when in 1793 a 'plaster' made of leather coated with 'unguentum hydrargyri Fortius' 'was apply'd thro mistake instead of another that had been ordered by the physician'.[55] When instructing the reader on making 'Stale Ale Mild & wholesome', the recipe advises that if 'half a Tea spoonful of salts of Tartar ... is not sufficient' then 'put a little more in till it ferments'. Thus, the judgement rested on the maker either observing signs of fermentation or knowing through experience the amount that would produce this effect.[56] Recipes usually assumed the reader had the requisite tacit knowledge to interpret and enact on the basis of limited details. As this recipe shows, the ability to extend the life of consumable goods was crucial in domestic settings with little access to refrigeration (besides a cool cellar) and a reliance on the seasonal production of everyday foodstuffs.

Whilst recipe books aimed to instruct, letters often revealed the anxieties of home producers about the efficacy of their methods. On 31 January 1701, Sir John Mordaunt wrote to his second wife Lady Penelope Mordaunt about her efforts to pickle pork. Pickling, as opposed to curing, meat was a popular method of preservation by the mid-eighteenth century and pork was the favoured vehicle, although mutton and beef were also treated in this way.[57] Pickled pork was referred to as 'tubbed bacon' on account of the technique of setting the meat in a dish to pickle and its associations with traditional cured meat. On this topic, Mordaunt worried that it was 'so fatt there is scares [scarce] any lean upon it', admitting that his investigation went only as far as looking 'upon yor Tubbd Bacon, but not tasted it'.[58] Mordaunt feared 'it is not done right' because he observed that the bacon was 'All drye except one of ye Hains at [the] bottome wch is cover'd wth Pickle'. Further, he concluded that the 'Tubb is too bigg for it' and he believed that 'it must be salted over again & ye Ham that was at yee Bottome laid at ye Topp'.[59] It is not clear whether John Mordaunt had any hands-on experience in pickling pork himself, but he felt able to identify possible errors in his wife's work. Nonetheless, the letter also communicated his optimism that the process could be repeated, with mistakes corrected and the potential for an edible result remained. In this way, through trial and error, observation and correction, serviceable 'tubb'd bacon' might be achieved.

A century later, in 1810, another household recipe book was begun by Mrs A. W. Baker of Ballytobin House, County Kilkenny in Ireland. Mrs Baker, like Mary Farewell, put particular store in knowledge that had been tried and tested by others. This example reveals that authority played a role within these texts and was highlighted for the reader. Baker's text makes frequent references to her grandmother's recipe book, referring the reader to that book for a method to 'Clear any Distilled Water that may be Muddy or Milky', and noting that a recipe for pickling walnuts was 'Strongly recommended in My Grandmother's Book'.[60] The trust put in her grandmother's methods helps us to understand a value system that emphasised authority through association with trusted sources and through personal experience. On the one hand, this process of recycling older recipes might indicate a conservatism within these documents, but recipe books also attest to a common understanding among their authors that local variables in equipment, ingredients, tastes, domestic affordance, climate and family needs all required consideration, experimentation and adaptation.

Baker's book, like many of its kind, includes a good range of pickling, preserving and potting recipes to make perishable goods serve year-round and she boasted that her recipe for 'Catsup [ketchup]' would 'last 20 years'.[61] For example, Baker's directions for pickling lemons advised that after quartering and salting, the lemons require a thorough drying out 'in the oven after the great heat is out, or in the sun'.[62] Ensuring the longevity of pickles and preserves was paramount, and the much-valued pickled walnut recipe suggested the double sealing of the 'pot' 'first with a Bladder, and outside that with Leather, that no air may get to them'.[63] Yet another pickling recipe noted the potential for cucumbers to masquerade as a more exotic ingredient: 'If you would have them taste like Indian mango put in Garlick instead of Shillots.'[64] Here, again, a preoccupation not only with the preservation of goods in the first instance but also with the revitalisation of a domestic product gone bad is clear. Moreover, the process of pickling was a creative one, allowing the cook not only to diversify her store cupboard year-round, but also to transform everyday ingredients into rare luxuries.

Mrs Baker, like Bishop Edward Synge, took an active interest in barm and her recipe book records six different entries on the subject. The importance of this material resource to the Baker home is further underlined by the Household Account Book for Ballytobin House, which identifies ownership of '1 Barm keg', revealing either specialised equipment for this purpose or – otherwise – the habitual use of a generic keg for barm, earning it that name.[65] Baker's book offers a range of options for preserving barm: for the usual week to ten days, for up to six weeks and a more elaborate method to preserve it for several months.[66] Her notes advise that it is possible to keep barm 'for Brewing without art' but for the purposes of baking, it required more care. However, the 'Method to Keep a Large Stock [of] Barm for either Bread or Cakes' marked a departure from the cycle of combining and cultivating mixtures of flour, water and barm and offered the baker something more akin to an active dry yeast.[67] As mentioned above, Baker discovered this method amongst the pages of a Dublin Society publication.

Throughout the recipe, Baker calls upon the reader to use their judgement and prior knowledge in deciding 'When you had good Barm a plenty' and determining what is 'a good Quantity' or when 'you have sufficient Quantity'.[68] However, the text also provides helpful descriptions of techniques; for example, in asking the reader to 'work it [the mixture] well with a Whisk', the recipe advises 'until it becomes liquid'. Essentially, thin layers of barm were painted with a brush onto the inner surface of a large tub or platter, which was then set upside down so 'that it May receive no dust but so that the Air may go under it to dry it'. The process was repeated until the layer was two or three inches thick and would 'serve for several months'.[69] Whilst devised for use in baking, this large stock of barm could still be applied to the purpose of brewing; in this case, the reader was instructed to cut off a piece, stir it into warm water and then 'take A large Handfull of Birch', dip it into the barm and hang it up to dry, taking care that 'no dust comes to it'. The next step is to 'Whissk it about in the Wort & then let it lye, when your drink works well take out the Broom & dry it & it will be fit for the Next

Brewing.'[70] This dried barm was clearly a versatile product that saved the baker or brewer some of the work of maintaining a live culture day-to-day. In her guidance on 'The French way of Making Leaven', Baker emphasised that 'From Xperience I know that six ounces of Leaven are not more than sufficient for A Quart of Flour', noting that 'If you are in a hurry or the Weather is cold you will require more.'[71] This comment demonstrates that the circumstances of a particular room were influenced by the environmental conditions of that region or season. Baker's book provides an exceptional level of guidance on matters of technique, but she ultimately confirms the importance of personal experience in the development of this tacit knowledge. The domestic records discussed here are unusually explicit in terms of technique and material knowledge and in this sense, like the Synge letters explored above, they are atypical archival survivals.

Weather diaries

Domestic habits of keeping a note extended well beyond the kitchen, and the recording of natural history was an increasingly widespread activity in eighteenth-century life. The greater accessibility afforded by the arrival of pre-formatted and printed journals for field notes in the 1770s was a significant factor in bringing a broader range of people to active natural history record-keeping. This development coincided with the sale of portable, pocket-sized guides to flora and fauna and the dissemination of 'user-friendly scientific classification systems' rendering a wider range of people capable of categorising and extrapolating from their own records.[72] It also reflected the increasing production in the later part of the century of small pre-formatted notebooks which were primarily used for the purposes of keeping notes of social engagements. In this way, wider trends in record-keeping and print culture are visible within the particular realm of natural history and reveal the connections between different cultures of journal-keeping.

Of course, there is no clear distinction between acts of record and those of observation, especially as they are made manifest in the archive. A handwritten list of birds sighted locally speaks not only of documentation but also of the act of live observation and the list itself represents a recognition of the relatedness of individual birds collected on the page and thereby categorised as a group. The keepers of natural history journals regarded the recording of things as a form of empiricism, a contribution to knowledge.[73] In these cases, making a record was both the means and the end. So here, it is proposed that habits of recording information about both domestic life and nature were overlapping and mutually supporting. Much like technique and tacit knowledge, they could be transferred and adapted from one context to the next. As such, practices of domestic record-keeping played an important role in shaping cultures of curiosity in this period.

Whilst natural historical record-keeping moved with the times, it also drew on a long history. Manuscript commonplace books, which prefigured other ephemeral formats such as the scrapbook, offered a place for individuals to compile knowledge.[74] Typical inclusions were extracts of favourite poems or proverbs, but commonplace books were heterogeneous by their nature, and their contents reflected the interests of the compiler. As such, commonplace books – like minds – were repositories of diverse kinds of knowledge, but they formed part of 'a pedagogic tradition related to rhetoric and the art of memory that dated back to the classical period'.[75] By the turn of the eighteenth century, common-placing was still being used as a form of information management that could facilitate the structuring of natural historical systems.[76] Thus, considerable overlap existed between ubiquitous forms of domestic record-keeping and scientific methods in this period – the latter drawing on characteristics of the former to concretise emerging systems of natural knowledge.

Here, the role of recording in histories of enquiry is discussed through two weather diaries. This was an enduring format of personal record-keeping and one that found expression in popular print culture. For many eighteenth-century observers, the weather revealed cosmic connections and formed one component part of an astrological worldview. For others, the weather brought signs of God's favour and displeasure and the documenting of its expression sat easily with the mode of providential accountancy and self-reckoning common to spiritual diaries of the early modern period.[77] More than anything else, the weather affected everybody and especially those whose livelihoods relied on crops. Careful record-keeping offered the individual the possibility of greater familiarity with the patterns of nature and, potentially, the elusive power of prediction. As Edward Synge, Bishop of Elphin, anxiously commented in a letter to his daughter on the subject of his crops, 'Nothing is certain, but uncertainty.'[78]

The personal weather diary had a significant relationship with periodicals like the *Gentleman's Magazine* and almanacs regularly carried weather-related features.[79] In the late seventeenth century, periodicals represented a quarter of all published titles and were a growing and accessible format of print.[80] Moreover, they relied heavily on reader response (whether real or fictional) as a means of generating interest in their varied offerings. Almanacs offered an incredible range of topics, including – among many others – information on the weather, astrology, astronomy, agriculture, tides and medical advice.[81]

Crucially, the weather diary was a format that could be read in a cheap almanac or the prestigious *Transactions of the Royal Society* and many outlets in between. The well-known cleric and naturalist, Gilbert White, had his own weather diary printed in the *Gentleman's Magazine*.[82] As 'B. M.' of Somerset wrote to the *Magazine* in late December 1781, the printing of 'Meteorological Journals ... were very much in repute', especially in the 1750s and 60s.[83] 'B. M.' insisted that the printing of such weather diaries sourced from across the country would reveal important data on 'the variation of the atmosphere' which, he argued, 'is certainly the principal cause of most of the epidemic diseases incident to our climate'. He mused upon whether 'experimental philosophers of this time think it beneath their notice to attend to such trivial matters' but submitted his own diary to the *Magazine* in the hopes that sense would prevail. Unfortunately, the editor noted that whilst they were grateful for the submission, its length and format (dissimilar to those 'kept at London') precluded its inclusion. Half a page was the most the *Magazine* could run to, but perfectly adequate for 'a comparison ... between two stations, and ... to judge the general state of weather in the southern part of the island'.[84] This episode reveals an active public discussion about the printing of these materials, an expectation of seeing them in outlets like the *Gentleman's Magazine* and a recognition of the collective nature of such record-keeping. The weather diary was a format that permeated eighteenth-century print culture whilst also representing a domestic, manuscript practice familiar to many. Whether conducted in pen and ink or perused in print, the weather and its patterns sparked widespread curiosity and diligent commitment to observation and record.

Here, Isaac Butler's manuscript Dublin weather diary (1716–34) and Richard Townley's 1791 publication *A journal kept in the Isle of Man, giving an account of the wind and weather, and daily occurrences, for upwards of eleven months: with observations on the soil, clime, and natural productions of that island* are examined up close. Butler and Townley's writings offer two views of the weather diary: Butler's manuscript is explicitly thus, containing nineteen years of near-complete daily weather records. Townley's book, on the other hand, is a publication based on a personal journal documenting the Isle of Man, but 'wind and weather' are the first in a sequence of subjects listed in the book's title. Both documents reference other occurrences, but Townley's piece encompasses a very diverse range of interests from natural phenomena to descriptions of cultural norms and man-made antiquities. Townley's inclusive approach was probably guided by the book market, where sellers aimed to offer something for every reader on their frontispieces, encouraging authors to cover an eclectic range of subjects.

Isaac Butler (*c.* 1691–1755) began his weather diary in 1716.[85] A Parish Constable of St Nicholas Without in Dublin, Butler was also a well-known almanac writer with encompassing interests in meteorology, botany, mineralogy, archaeology and astrology.[86] Unsurprisingly for a compiler of almanacs, Butler relied heavily on astrological explanation and his diary notes relevant details about the status of the moon and the position of planets – making connections between celestial bodies and the patterns of terrestrial weather.[87] However, Butler's interests in the natural sciences and antiquity were furthered in his role as an 'inquirer' for the Physico-Historical Society, for whom he collected and reported evidence on a variety of subjects concerning Ireland.[88] In its inaugural year of 1744, the Society noted Butler's efforts travelling through the counties of Dublin, Meath, Westmeath, Longford and Louth where he mapped the local geography, observed and collected samples of 'diverse rare plants' and compiled information regarding local historical artefacts, buildings and fossils.[89] These activities and institutional affiliation suggest that Butler's weather diary formed one part of a wider set of intellectual commitments.

Butler was not alone in this endeavour; a rich culture of weather observation existed in Ireland with at least 750 records being taken before 1850, some under the auspices of societies and many others taken by individuals across the island.[90] The vast majority were observations taken without the benefit of an instrument of measurement and records were reproduced in the pages of Irish societies' publications alongside cheaper print productions.[91]

Butler's vocabulary for describing the weather was applied consistently over time. For example, sunshine was 'fair', 'pleasant', 'hot' and sometimes 'serene'. Rain could be 'misling', 'driving', 'dropping' or come in 'showers'. Alongside the commentary on daily weather conditions, there are also notes on the moon's status, such as it being in the 'latter degrees of Scorpio', which was connected to the likelihood of 'Clouds and rain'. The planets and astrological signs attracted plenty of mentions, such as the 'sextile of Jupiter and Venus' and 'Saturn and Mars in opposition from Libra to Aries'. The integration of terrestrial and celestial observations and the explanatory power of astrology are key features, revealing the longevity of these ideas among educated civic officials in this period.

Another inclusion is a monthly record of unusual occurrences in Ireland or from around the world. For example, in June 1717, Butler noted that a mountain 'beyond Rathfarnum [Rathfarnham]' in Co. Wicklow had 'bursted open from whence issued a great irruption of Waters with a prodigious Noise, it bore down stones of incredible Bulk, and form'd in a Valey beneath a piece of Natural pavement not to be parallel'd'.[92] In December 1717, a series of dramatic weather events were recorded across the continent, which became known as the Christmas Flood of 1717. The diary recorded '2500 bodies of persons drowned' in the Dutch city of Groningen, adding, 'Melancholy are the Accounts from the North parts of Holland and Germany of the great Damages and Losses sustain'd there by the great storms.'[93] These monthly entries did not confine themselves to weather-related occurrences, noting other phenomena including political events.

On 19 February 1719, Butler noted that 'A Meteor or Ball of fire' was sighted.[94] This focus on celestial matters was, of course, in keeping with his astrological perspective. The meteor appeared 'southward of Dublin its altitude did not exceed 13 deg'. And it 'gave so great a light as to efface that of the Moon and stars which then shone'; moreover, 'in its progressive motion from east to west it left a great train of smoak behind it, and all of a sudden Extinguish'd'.[95] Butler noted that the same meteor was seen in Paris, but other records from the period reveal that astronomers in Bologna also recorded its occurrence and attempted to measure its altitude, describing the meteor as the same size as the moon and as bright as the sun near the horizon and throwing out sparks and smoke.[96] The fact that Butler made mention of a measurement of the altitude suggests that he either accessed information on astronomy through Dublin contacts or print culture or that he made the measurement himself. In Chapter 4, a community of Dublin astronomers operating in the mid-century will be discussed in more detail, revealing the accessible nature of this scientific pursuit at this time.

The second diarykeeper, Richard Townley, served as a sheriff of Lancashire from 1752 to 1753 and died at Ambleside in 1802. The family were of the gentry class and the *Journal* speaks to the kind of leisured life that could accommodate a long-term enquiry pursued away from home. Townley's two-volume *Journal* made an account of the wind and weather as part of a broader exploration of the natural and cultural features of the Isle of Man. As a visitor to the island, Townley drew on traditions of travel writing and the practices promoted by intellectual societies that sought to map, record, collect and capture the important natural historical and antiquarian characteristics of different regions.

Townley offered the reader 'an easy, *novel* mode of information' about 'a *sister* island; being now, in great part, a member of the British Empire'.[97] He promoted his publication as the product of an informal process of information-gathering as opposed to a '*formal* history' that might have required a more structured approach. He urged, therefore, that his readers 'must not expect, accuracy either in design, method, or language; but a mere piece of patchwork'.[98] The *Gentleman's Magazine* 'Review of publications' was unimpressed with the *Journal*, describing it as 'a dull journal of uninteresting events, intermixed with a meteorological diary, and interlaided thick with hackneyed quotations'.[99]

The vast majority of entries begin with a comment on the weather. For example, 'a fair morning' in June offered a sought-after chance for 'a scramble amongst the rocks, round Douglas-head', and Townley 'judged this morning, as free from fog and any appearance of rain, a very favourable opportunity for undertaking it'.[100] However, the weather often comes between Townley and his exploration of the Isle, such as on 6 June 1789 when he 'was obliged to turn about, and make a retreat from my favourite walk in order to preserve my eyes from the sharp sand, that was driven so furiously against them'.[101]

The *Journal* documents many different facets of island life, but with a substantial local fishing industry, sea life looms large in this account with regular references to the prospects of the fishing boats, the daily catch and the fluctuating price of fish and other commodities at market.[102] Observations of nature are often driven by an interest in its value from a subsistence or commercial perspective. For example, a boat trip to Calf Island revealed a colony of 'sea-parrots' (puffins). When treated to 'a dish of cold parrots' by the 'very civil old lady' who lived with her husband as the only inhabitants of the island, Townley found them 'uncommonly good and nourishing'.[103]

On another occasion, when observing the breeding habits of local herring shoals, Townley chooses to believe the evidence of his own eyes over the descriptions given by experts. On arrival on the Isle of Man, Townley had 'received with great caution' the stories of herring breeding 'in the different bays and channels about the island' on account of having been informed 'by naturalists, and other writers upon the subject' that this fish 'bred in very *high* northern latitudes'.[104] In fact, in 1786, the *Transactions of the American Philosophical Society* published an article by a Mr John Gilpin in which he acknowledged that the location of spawning 'remains a query for naturalists' and made the argument for the 'bays, rivers, creeks, and even small streams' of New England.[105] However, Townley adjusted his view when 'two or three times' he 'had an opportunity of seeing, and observing, considerable shoals of young herring, that, from their diminutive size could never have journeyed from shores very distant from *this* small island'.[106] Here it is clear that Townley had not only engaged with published natural historical writings but was willing to diverge from the view of specialists when he could see the evidence for himself.

Like other keepers of nature journals, Townley adhered to an ethos that recognised both the value of observing and recording but also the partial nature of that record. On 15 June 1789, he reflected on this point, noting that 'Wise and humble men' will 'confess their inability to develop the secret workings of nature which do, and ever will, serve to puzzle and confound the most improved understandings, thereby effectually humbling the pride of science'.[107] However, it is in these paragraphs referring to the unknowability of God's creation that Townley delivers some of his most detailed observations, in this case of seaweed.

One 'species of sea-weed' is introduced in terms of the 'most wonderful manner' in which it arrives on the beaches of one of the island's bays 'constantly at this season of the year', 'dragging after it a stone of *forty*, *fifty*, or (perhaps) *sixty* times its own weight'.[108] The grip of the seaweed is articulated with care: 'They are furnished with roots, or rather stoles', and when first washed ashore 'and are fresh and vigorous, there is no separating them from their new-adopted friends, but by acts of strong violence'.[109] Townley appears to have tested the strength of the seaweed's attachment to the rocks in various ways: 'I have several times brought a stone along with me, of five or six pounds weight, for many hundred yards, by the adhesion alone of a plant not two ounces in weight.'[110] The description of the attachment is unusually thorough by the standards of the *Journal*, and uses close observation ('From those stoles, several short crooked roots branch out on every side, resembling the hooks of the ivy, and other creeping plants') and everyday examples to elaborate the physical features of the seaweed's roots ('exactly resembling those *pliant* pieces of leather, which boys use as *take-ups*').[111] Seaweed was a troublesome object of enquiry in this period, on account of the difficulty of observing it underwater and its fast deterioration once washed up on shore.[112] Despite Townley's commitment to observation as opposed to analysis of natural phenomena, his entry on seaweed offers some supposition:

> The only conjecture I can make, respecting the wonderful appearance, is, That those plants, being torn up from their *native* beds (within the great deeps of the ocean) by furious winds and waves, are carried along by the tide current ... till they arrive in shallow water; where, being allowed a little rest they *instinctively* form those new connections, embracing and clinging to the stones as anchors of safety, such as will prevent them being entirely driven out of their *own natural* element.[113]

Here, he offered something more than a record and demonstrated a confidence developed through first-hand experience.

Townley positioned himself as an inquisitive observer who took up the concerns of the naturalist or antiquary but who did not claim that title. However, he held a sceptical view of travel writing as a reliable source of information. For example, he took issue with 'Voyagers, to various parts of the world' who 'speak very confidently of *flying-fish*'.[114] In Townley's 'own poor opinion upon the subject', it seems that on the contrary, fish are merely jumping out of the water.[115] As something of a travel writer himself, this disquisition on truth-telling and the recording of natural history underlines the ambiguity of Townley's own text – which regularly shifts in style between a range of genres of journal.

Although Townley's published journal is not principally a weather diary, it uses the rhythms of that format to anchor its lengthy daily narratives and invokes several other diary forms alongside. This hybrid text illuminates the comfortable familiarity of the author with a range of genres that were commonly found in manuscripts and print in this period. Potential diary writers did not need the example of Fellows of the Royal Society as an inspiration; they needed only draw upon the facets of their own local and domestic environment to take pen in hand. Moreover, the search for patterns and predictability which were inherent in many of these forms of record-keeping had a strong relationship with the power and control offered by household accounting and also with modes of autobiographical or life-writing that were also common in this period. In this way, eighteenth-century individuals might easily keep account, simultaneously, of themselves and their environments – taking practices, habits and motivation from one written form to the next.

Conclusion

The record-keepers and letter-writers discussed in this chapter offer a glimpse of the range of skills and tacit knowledge needed to run and provision a home in this period. Account books and inventories speak to the material and spatial affordances of households, but recipe books and letters illuminate specific features of material knowledge and also the networks of social connection and rhythms of repetition that allowed that knowledge to be accessed, exchanged and, ultimately, trusted. An empiricism forged in record-keeping blossomed in the eighteenth century.[116] By exploring how a variety of domestic processes conditioned and extended understanding, it is possible to see how knowledge worked through practices and also how experience could be converted into evidence.

Whilst historical analysis of the home has incorporated a sense that household accounting was highly prevalent in the eighteenth century, acts of keeping a note are not always viewed as specifically domestic in character. Both weather diaries discussed here show their authors' aspirations for knowledge creation through record-keeping, but they also reveal the influences of contemporary domestic habits of keeping notes and older traditions of common-placing, diary and travel writing. Given the sheer quantity and range of household record-keeping at this time and individuals' familiarity with these forms and modes, this chapter has argued that record-keeping was implicitly domestic in character and that the demands of home shaped these written practices as they were put to the service of science. The next three chapters shift focus to examine specific domestic practices themselves (collecting, observing and experimenting) – practices that this book argues were rooted in the domestic environment in this period.

Notes

1 Timothy J. Clark, *The sight of death: An experiment in art writing* (New Haven, CT and London: Yale University Press, 2006), p. 9. Nicolas Poussin (1594–1665), 'Landscape with a man killed by a snake' (probably 1648), The National Gallery, NG5763; 'Landscape with a calm' (1650–51), Getty Museum, 97.PA.60.

2 Polanyi, *Tacit dimension*; Michael Polanyi, *Personal knowledge: Towards a post-critical philosophy* (New York: Harper & Row, 1964), p. 50; Mark T. Mitchell, *Michael Polanyi: The art of knowing* (Wilmington, DE: ISI Books, 2006), pp. 59–103 (p. 63).

3 See for example, Pamela H. Smith (ed.), *Entangled itineraries: Materials, practices, and knowledges across Eurasia* (Pittsburgh, PA: University of Pittsburgh Press, 2019); Lissa Roberts, Simon Schaffer and Peter Dear (eds), *The mindful hand: Inquiry and invention from the late Renaissance to early industrialisation* (Amsterdam: Edita KNAW,

2007); Marieke Hendriksen, '"Art and technique always balance the scale": German philosophies of sensory perception, taste, and art criticism, and the rise of the term technik, ca. 1735–ca. 1835', *History of Humanities*, 2:1 (2017), pp. 201–19; Matteo Valleriani (ed.), *The structures of practical knowledge* (Cham, Switzerland: Springer, 2017).

4 Pamela O. Long, *Artisan/practitioners and the rise of the new sciences, 1400–1600* (Corvallis, OR: Oregon State University Press, 2014), pp. 10–29.

5 Marieke Hendriksen, 'Review of *The structures of practical knowledge*, edited by Matteo Valleriani', *Ambix*, 66:1 (2019), pp. 88–90; H. Otto Sibum reconceives Polanyi's understanding of tacit knowledge 'as the expression of a historically located gestural knowledge' which escapes historical attention because it belongs 'to different worlds of sense', 'Science and the knowing body: Making sense of embodied knowledge in scientific experiment' in Sven Dupré, Anna Harris, Julia Kursell, Patricia Lulof and Maartje Stols-Wilcox (eds), *Reconstruction, replication and re-enactment in the humanities and social sciences* (Amsterdam: Amsterdam University Press, 2020), p. 285 (pp. 275–93).

6 Havard, 'Preserve or perish', pp. 3, 20. Using reconstruction as a method, Havard also makes the case for preservation as a form of experiment – emphasising its demand for close observation, trying and testing and deciding on the relative efficacy of different methods. Pennell, *English kitchen*; Anita Guerrini, 'The ghastly kitchen', *History of Science*, 54:1 (2016), p. 92 (pp. 71–97).

7 Elaine Leong and Alisha Rankin, 'Testing drugs and trying cures: Experiment and medicine in medieval and early modern Europe', *Bulletin of the History of Medicine*, 91:2 (2017), p. 181 (pp. 157–82).

8 See, for example, Easterby-Smith, 'Recalcitrant seeds', pp. 235–6.

9 John R. Harris, 'Skills, coal and British industry in the eighteenth century', *History*, 61 (1976), pp. 167–82.

10 See, for example, 'of the leaven made of Potatoes' in *Transactions of the Dublin Society*, vol. 1, pt. 2 (1799), pp. 60–1; 'Receipt for potatoe barm' in *Transactions of the Dublin Society*, vol. 2, pt. 1 (1800), pp. 351–2; 'A substitute for barm' in *Transactions of the Dublin Society*, vol. 2, pt. 2 (1801), p. 279; and NLI, 'Mrs. A. W. Baker's Cookery Book, vol. 1, 1810', MS 34,952, fos 34–5.

11 David Hayton, 'Review: Marie-Louise Legg (ed.), *The Synge letters: Bishop Edward Synge to his daughter Alicia: Roscommon to Dublin, 1746–1752* (Dublin: Lilliput Press, 1996)', *Irish Historical Studies*, 30:119 (1997), pp. 479–80.

12 Marie-Louise Legg (ed.), *The Synge letters: Bishop Edward Synge to his daughter Alicia, Roscommon to Dublin 1746–1752* (Dublin: Lilliput Press, 1996): Synge to daughter, Elphin, 16 Jul. 1751, p. 324.

13 The palace was a palladian-style mansion attributed to the Dublin architect Michael Wills and was built in 1747–49; see also Hayton, 'Review'.
14 Legg, *Synge letters*, p. 325: Edward Synge to Alicia Synge, 16 Jul. 1751.
15 *Ibid.*; winters were spent in the city and away from the Roscommon estate where Jane baked her bread.
16 Legg, *Synge letters*, pp. 325–6: Edward Synge to Alicia Synge, 16 Jul. 1751.
17 Legg, *Synge letters*, p. 326: Edward Synge to Alicia Synge, 16 Jul. 1751.
18 Legg, *Synge letters*, p. 325: Edward Synge to Alicia Synge, 16 Jul. 1751.
19 *Ibid.*
20 *Ibid.*; the term 'pitch' seems to be used to mean 'settle', it appears many times in the lengthy descriptions of the process in these letters; this term is used in the same way in some parts of the West Country in England and applied to snow when it settles.
21 Legg, *Synge letters*, p. 326: Edward Synge to Alicia Synge, 16 Jul. 1751.
22 See Thirsk, *Food*, pp. 232–3.
23 Legg, *Synge letters*, p. 326: Edward Synge to Alicia Synge, 16 Jul. 1751.
24 *Ibid.*
25 *Ibid.*
26 *Ibid.*
27 *Ibid.*; 'pitch' meaning settle.
28 Legg, *Synge letters*, p. 331: Edward Synge to Alicia Synge, 23 Jul. 1751.
29 Legg, *Synge letters*, p. 326: Edward Synge to Alicia Synge, 16 Jul. 1751.
30 Legg, *Synge letters*, p. 331: Edward Synge to Alicia Synge, 23 Jul. 1751.
31 Wormwood and artichoke leaves were both ingredients used to aid digestion in this period, the former being used also as a purging remedy; see Anne Stobart, *Household medicine in seventeenth-century England* (London: Bloomsbury, 2016), p. 94. Although wormwood is also famous for its use in absinthe, this development takes place at the end of the eighteenth century in Switzerland, so is unlikely to be relevant to this mid-century Irish example.
32 Legg, *Synge letters*, p. 325: Edward Synge to Alicia Synge, 16 Jul. 1751.
33 Legg, *Synge letters*, pp. 326, 331: Edward Synge to Alicia Synge, 16 Jul. 1751; 23 Jul. 1751.

34 Legg, *Synge letters*, p. 331: Edward Synge to Alicia Synge, 23 Jul. 1751.
35 *Ibid.*
36 *Ibid.*
37 *Ibid.*
38 Legg, *Synge letters*, p. 326: Edward Synge to Alicia Synge, 16 Jul. 1751.
39 *Ibid.*
40 *Ibid.*
41 *Ibid.*
42 Legg, *Synge letters*, p. 331: Edward Synge to Alicia Synge, 23 Jul. 1751.
43 *Ibid.*
44 *Ibid.*
45 *Ibid.*
46 Jason Scott-Warren, 'Early modern bookkeeping and life-writing revisited: Accounting for Richard Stonley', *Past & Present*, 230:11 (2016), pp. 151–70.
47 Harvey, *Little Republic.*
48 Leong, *Recipes*; Sally A. Osborn, 'The role of domestic knowledge in an era of professionalisation: Eighteenth-century manuscript medical recipe collections' (PhD thesis, University of Roehampton, 2016).
49 Derbyshire Record Office (hereafter DRO), Wright of Eyam Hall, D5430/50/4; it has not been possible to precisely identify Mary Farewell, but a Jane Farewell (1700–74) married John Wright (1700–80), grandson of the John Wright who built Eyam Hall in Derbyshire.
50 DRO, Wright of Eyam Hall, D5430/50/4.
51 *Ibid.*
52 Herbert, *Female alliances*, pp. 78–116, although recipes for medicinal remedies are often attributed to a male doctor and some recipe collectors harvested widely. See for example Carolyn Powys's recipe book, BL, 'Powys diaries', Add MS 42173; and Appendix 4 in Osborn, 'Role of domestic knowledge'.
53 DRO, Wright of Eyam Hall, D5430/50/5.
54 *Ibid.* Joan Thirsk has commented on the lively recipe-related discourse taking place in eighteenth-century newspapers; see *Food*, pp. 158–9.
55 DRO, Wright of Eyam Hall, D5430/50/5; 'unguentum hydrargyri Fortius' is a strong mercury-based ointment.
56 DRO, Wright of Eyam Hall, D5430/50/5.
57 Thirsk, *Food*, pp. 133, 169–70.
58 WCRO, Mordaunt Family of Walton, CR1368/Vol. 1/11: John Mordaunt to Penelope Mordaunt, 31 Jan. 1700/1.

59 *Ibid.*
60 NLI, 'Mrs A. W. Baker's cookery book, vol. 1, 1810', MS 34,952, fos 7; 9–10.
61 *Ibid.*, fo. 38; the Irish Quaker and diarist, Mary Leadbetter, likewise referred to making fruit preserves, alcohol and vinegar at home and the seasonal quality of these activities according to periods of fruit-picking; McLoughlin, 'Sober duties of life', pp. 77–8.
62 NLI, 'Mrs A. W. Baker's cookery book, vol. 1, 1810', MS 34,952, fo. 10.
63 *Ibid.*
64 *Ibid.*, fo. 35.
65 NLI, 'Household account book, 1797–1832', MS 42,007, fo. 57; this book most likely belonged initially to Mrs Baker herself but was later kept by her daughter-in-law Charity Baker.
66 NLI, MS 34,952, fos 17, 34–5.
67 When brewing moved from top-fermenting to bottom-fermenting yeast in the nineteenth century, new methods to generate yeast for baking purposes were developed, culminating in a dry baker's yeast *Saccharomyces cerevisiae* that is still used today.
68 NLI, MS 34,952, fo. 34.
69 *Ibid.*
70 *Ibid.*, fos 34–5; 'wort' refers to the infusion of malt prior to fermentation in the brewing of beer.
71 *Ibid.*, fo. 19.
72 Mary E. Bellanca, *Daybooks of discovery: Nature diaries in Britain, 1770–1870* (London: University of Virginia Press, 2007), pp. 11–12; she notes that this form of nature journal grew out of older practices of diary-keeping and life-writing but converted them into being explicitly knowledge-seeking; for the histories of note-taking that facilitated science also see Daston, 'Empire of observation', pp. 95–9.
73 Bellanca, *Daybooks of discovery*, p. 15.
74 See, for example, Ann Blair, *The theater of nature: Jean Bodin and Renaissance science* (Princeton, NJ: Princeton University Press, 1997) and Ann Moss, *Printed commonplace-books and the structuring of Renaissance thought* (Oxford: Oxford University Press, 1996).
75 Lucia Dacome, 'Noting the mind: Commonplace books and the pursuit of the self in eighteenth-century Britain', *Journal of the History of Ideas*, 65:4 (2004), p. 603 (pp. 603–25); see also David Allen, *Commonplace books and reading in Georgian England* (Cambridge: Cambridge University Press, 2010).
76 M. D. Eddy, 'Tools for reordering: Commonplacing and the space of words in Linnaeus's *Philosophia Botanica*', *Intellectual History*

Review, 20:2 (2010), pp. 227–52; see also Roberts and Werrett, *Compound histories*, p. 26 for discussion of the increasing 'governance through paper' that shaped eighteenth-century chemistry and society.

77 Golinski, *British weather*, p. 81; Raymond Gillespie, 'Climate, weather and social change in seventeenth-century Ireland', *Proceedings of the Royal Irish Academy: Archaeology, Culture, History, Literature*, 120C (2020), pp. 263–71; James Kelly, 'Climate, weather and society in Ireland in the long eighteenth-century: The experience of the later stages of the Little Ice Age', *Proceedings of the Royal Irish Academy: Archaeology, Culture, History, Literature*, 120C (2020), pp. 273–324; see also Effie Botonaki, 'Seventeenth-century English women's spiritual diaries: Self-examination, covenanting, and account keeping', *The Sixteenth Century Journal*, 30:1 (1999), pp. 3–21.

78 Legg, *Synge letters*, p. 324: Edward Synge to Alicia Synge, 16 Jul. 1751.

79 See, for example, *Gentleman's Magazine*, vol. 24 (Feb. 1754), pp. 58–9; vol. 24 (Mar. 1754), p. 106; vol. 24 (Apr. 1754), pp. 151–2, which includes comparisons between places and approaches.

80 Berry, *Gender, society and print culture*, pp. 17–18.

81 Bernard Capp, *Astrology and the popular press: English almanacs, 1500–1800* (London: Faber, 1979); R. C. Simmons, 'ABCs, almanacs, ballads, chapbooks, popular piety and textbooks' in John Barnard and D. F. McKenzie (eds), *The Cambridge history of the book in Britain*, vol. 4 (Cambridge: Cambridge University Press, 2014), pp. 504–13; the Stationers' Company enjoyed exclusive rights over the publishing of almanacs from the early seventeenth century through till a court case in 1775: 'Stationers' Company v Carnan'; production peaked at the turn of the eighteenth century when between 350,000 and 400,000 almanacs were being printed in the last two months of every year, see Louise Hill Curth, 'The medical content of English almanacs, 1640–1700', *Journal of the History of Medicine*, 60:3 (2005), p. 258 (pp. 255–82).

82 Golinski, *British weather*, pp. 56, 65–7.

83 *Gentleman's Magazine*, vol. 52 (Feb. 1782), p. 65.

84 *Ibid.*; half a page was exactly the space offered to a meteorological diary covering March 1781, featured in the same edition, p. 51.

85 Dublin City Library and Archive (hereafter DCLA), Gilbert Collection, MS 132: Isaac Butler, 'The diary of weather and winds. For 19 years commencing with AD 1716 and concluding with 1734. Exactly observed and taken at the City of Dublin'; Butler's diary is thus one of the earliest daily records of weather in Ireland and the period he covered witnessed comparatively warm weather, especially as compared

with the cold winters of the 1690s and the colder conditions experienced from the 1740s onwards; see Michael G. Sanderson, 'Daily weather in Dublin 1716–1734: The diary of Isaac Butler', *Weather*, 73:6 (2018), p. 179 (pp. 179–82).

86 C. J. Woods, 'Butler, Isaac', *Dictionary of Irish biography* (2009): https://doi.org/10.3318/dib.001249.v1; Mary Pollard, *A dictionary of members of the Dublin book trade, 1550–1800* (London: Bibliographical Society, 2000), pp. 68, 603–4.

87 See Eoin Magennis, '"A land of milk and honey": The Physico-Historical Society, improvement and the surveys of mid-eighteenth-century Ireland', *Proceedings of the Royal Irish Academy: Archaeology, Culture, History, Literature*, 102C:6 (2002), p. 205 (pp. 199–217).

88 The Physico-Historical Society was short-lived (1744–52) but achieved a great deal, including county surveys for Down, Waterford, Cork, Kerry and Dublin, and the dissemination of research into Irish coinage and mineral water among other topics, see Magennis, 'Land of milk and honey', p. 200; Gordon L. Herries Davies, 'The Physico-Historical Society of Ireland, 1744–1752', *Irish Geography*, xii (1979), pp. 92–8.

89 See Magennis, 'Land of milk and honey', p. 205.

90 Carla Mateus, 'Searching for historical meteorological observations on the island of Ireland', *Weather*, 76:5 (2021), p. 161 (pp. 160–5); organisations orchestrating the collection of meteorological information include the Royal Irish Academy, Dublin Philosophical Society, the Royal Dublin Society, the Royal Cork Institution and the Kilkenny Society; see Mateus, 'Searching', 162.

91 Society publications include the *Philosophical Transactions of the Royal Society of London*; *Proceedings of the Royal Irish Academy*; *Transactions of the Royal Irish Academy*; *The Proceedings of the Royal Dublin Society*; *The Journal of the Royal Dublin Society* and *The Scientific Proceedings of the Royal Dublin Society*; other magazines include the *Annals of Philosophy, or, Magazine of Chemistry, Mineralogy, Mechanics, Natural History, Agriculture, and the Arts*; *The Irish Farmer's and Gardener's Magazine and Register of Rural Affairs*; and *The Munster Farmer's Magazine, Conducted under the Direction of a Committee of the Cork Institution*; see Mateus, 'Searching', p. 162.

92 DCLA, Gilbert Collection, MS 132, fo. 36: entry for Jun. 1717.

93 *Ibid.*, fo. 46: entry for Dec. 1717.

94 *Ibid.*, fo. 76: entry for Feb. 1719.

95 *Ibid.*

96 P. M. Millman, 'Meteor news – telescopic meteor observations; the Ierofeevka Meteorite; meteor heights determined in the XVIII century', *Journal of the Royal Astronomical Society of Canada*, 31 (1937), pp. 364–5 (pp. 363–6).

97 Richard Townley, *A journal kept in the Isle of Man*, vol. 1 (Whitehaven, 1791), pp. xiii, xvi; italicisation from the original.
98 Townley, *Journal*, vol. 1, p. xiii.
99 *Gentleman's Magazine*, vol. 62, pt. 2 (1791), p. 840.
100 Townley, *Journal*, vol. 1, p. 52: entry for 9 Jun. 1789.
101 Townley, *Journal*, vol. 1, p. 50: entry for 6 Jun. 1789.
102 Townley, *Journal*, vol. 1, e.g. pp. 91, 93, 107, 147.
103 Townley, *Journal*, vol. 1, pp. 55, 58, 62: entries for 11 and 13 Jun. 1789.
104 Townley, *Journal*, vol. 1, p. 88: entry for 5 Jul. 1789.
105 J. Gilpin, 'Observations on the annual passage of herrings', *Transactions of the American Philosophical Society*, vol. 2 (1786), pp. 236–9.
106 Townley, *Journal*, vol. 1, pp. 88–9: entry for 5 Jul. 1789.
107 Townley, *Journal*, vol. 1, p. 64: entry for 15 Jun. 1789.
108 Townley, *Journal*, vol. 1, pp. 64–5: entry for 15 Jun. 1789.
109 Townley, *Journal*, vol. 1, p. 65: entry for 15 Jun. 1789.
110 *Ibid.*
111 *Ibid.*
112 Anne Secord, 'Coming to attention: A commonwealth of observers during the Napoleonic Wars' in Daston and Lunbeck, *Scientific observation*, pp. 426–7 (pp. 421–44).
113 Townley, *Journal*, vol. 1, p. 66: entry for 15 Jun. 1789.
114 Townley, *Journal*, vol. 1, p. 109: entry for 15 Jul. 1789.
115 *Ibid.*
116 Barbara M. Benedict, 'Collecting trouble: Sir Hans Sloane's literary reputation in eighteenth-century Britain', *Eighteenth-Century Life*, 36:2 (2012), pp. 115–16 (pp. 111–42).

Part II

3

Collecting

Material accumulation is a common enough human instinct, although the scope to indulge varies widely. For those with the least in the eighteenth-century British world, the ability to collect might extend as far as a locked box or even just a pocket. For others, large households with room upon room were the compass of their material accretion. When domestic accumulation is termed 'collecting', it takes on rather grand connotations – imperial trade, global travel and the big names of Britain's earliest museums: Tradescant and Sloane.[1] Moreover, the term 'collecting' implies a highly conscious approach with a view to developing and sharing personal curiosity or furthering knowledge. Nonetheless, collecting things at home was an activity that anyone with a little disposable income could do, and many did. Some people styled household accumulations of one variety or another as a conscious 'collection'; others just amassed according to their own interest, use or desire. Whilst eighteenth-century readers of the periodical press may have marvelled at the findings of naturalists, imperialists and voyagers, they would not have been surprised to hear about the methods they used. These methods – record-keeping, letter-writing, journal-keeping, collecting, drawing and annotating – were wholly familiar.

In this period, the mind itself was often characterised as a collection of accumulated artefacts. As Sean Silver has discussed, the view of the imagination as a creative force was the product of a later age and, in the eighteenth century, the mind was much more likely to be conceptualised as a storehouse of objects, ordered and assessed through wit and reason.[2] The analogy of the collection was pervasive and also found expression in print culture, most especially in the periodical press. Even the word 'magazine' was understood as a repository of sorts and acted as a synonym for museum.[3] The notion of the mind as a collection had a long history, recognisable in the works of Aristotle, Renaissance humanism and seventeenth-century natural philosophy. However, the concept gained renewed cultural purchase in the eighteenth century.

This was also an era in which museums proliferated and became increasingly accessible to non-elite viewers.[4] Famously, the British Museum opened semi-publicly in 1753, but other collections in private houses and urban coffeehouses contributed to the opportunities for viewing antiquities, natural specimens and hybrid assemblages. For example, Don Saltero's coffeehouse on Cheyne Walk in Chelsea was one of London's most notable attractions and had been established in the late seventeenth century by a servant of Sir Hans Sloane, James Salter.[5] Don Saltero's housed over 10,000 artefacts and specimens, which visitors could inspect when they dropped by for refreshments, a haircut or even dental work.[6] Sir Ashton Lever (1729–88) brought his collection of around 27,000 objects from Alkrington Hall in Lancashire to Leicester Fields (later Leicester Square) in London in 1775, charging a fee of five shillings and three pence for entry.[7] Later in the period, William Bullock (*c.* 1773–1849) moved a 32,000 object collection from Liverpool to Piccadilly, London, including natural history, archaeology and ethnography.[8] Irish physicians (Sloane being a famous example) were active in epistolary networks and associational activities that saw them collecting natural history and *materia medica*.[9] Small collections were visible in the homes of the middling sort upwards and other social spaces like public houses and coffeehouses, especially in towns and cities.

Many of the earliest collections that became publicly accessible in London had of course started life as private collections in homes, which invited visitors were able to view. Even when such collections inhabited new walls, they no doubt bore the marks of their domestic origins in the quantities and kinds of objects that had been acquired and which originally fitted into the spatial parameters of a given household. Ashton Lever's Alkrington Hall was one such household built on a large scale: a three-storey, brick mansion in the Classical style designed for the Lever family and erected in the 1730s. However, much more modest collections were to be found in many homes in this period. For example, on 15 January

1829, Benjamin Williams and John Brinkley were tried at the Old Bailey for breaking into the stationer Michael Watson's home on 26 November 1828 and stealing a range of objects including two vases, pasteboard ornaments, a pheasant's tail and five shells among other items of value.[10] According to Watson's servant, Susannah Stracey, the items were stolen from the cabinet of curiosities and the mantelpiece in the 'summer room', which overlooked the river Thames at this Wapping address.[11] In this case, a witness who lived in Chancery Lane and dealt in 'shells and other articles' attested to Brinkley later selling him some similar items for six or seven shillings.[12]

Along the same lines, on 20 October 1784, Robert Artz and Thomas Gore stood accused of shoplifting from Hyam Hart's shop in Hemmings Row in London's West End.[13] Hyam kept 'a shop of curiosities of all kinds, pictures, shells, fossils'.[14] On account of Hart being 'very often at sales' acquiring merchandise for his 'cabinet of curiosities', his wife Deborah Hart and her son were in the shop at the time of the theft.[15] In these accounts, it becomes clear that 'cabinets of curiosity' were present in a wide range of establishments, including modest homes, shops and public spaces of leisure such as coffeehouses. As such, those who collected and those who could visit a collection were a growing population. As strong connections were drawn between observing 'curiosities' and understanding nature, the increasing accessibility of natural historical specimens to urban residents who might lack the ability to collect from nature themselves was a significant development.

Whilst homes often accommodated collections, networks of sociability and business procured the objects themselves. Middling sort collectors could reach a wide range of spaces and groups for the acquisition of specimens.[16] Studies of Sir Hans Sloane's acquisitions reveal the network of other collectors he relied upon, including entomologist Eleanor Glanville (1654–1709); the famous naturalist and apothecary James Petiver (*c.* 1655–1718); and shoe-maker and book-seller John Bagford (1650–1716).[17] As mentioned above, it was Sloane's own servant James Salter who established and ran one of London's best-known coffeehouse museums.

As much as social connection facilitated collecting, the practice was also firmly linked with another mainstay of the domestic and commercial world: record-keeping. This was true of collectors up and down the social scale. Hans Sloane used detailed record-keeping about the circumstances and encounters that brought an object into his possession, and that object's own history of ownership, in order to develop 'a new form of collecting'. By doing so, Sloane 'promoted the practice as a method of understanding and classifying experience itself and turned his curiosity cabinet into a kind of university'.[18] Sloane was not alone in expanding the ambitions of collecting in this period. The Duchess of Portland established a collection that combined diverse priorities, including the extension of understanding through categorisation, aesthetic presentation, entertainment, sociability, wonder and delight.[19]

Alongside the positive images of men and women of science – or *virtuosi* – discovering the secrets of nature, elite habits of collecting were also often satirised as compulsive, greedy and obsessive and collecting material things at any strata of society certainly posed questions concerning the prudence of the purchaser. However, curiosity and the urge to amass clearly overcame such social censure as eighteenth-century homes readily filled up with objects and specimens. What follows explores the domestic activity of collecting, revealing both the ubiquity and diversity of this practice in the period. In doing so, this chapter (alongside Chapters 4 and 5) seeks to undermine a view of science as a series of rarefied practices undertaken by scholars who then shared their findings with the 'public' via print and suggests, instead, that the actions of scientists were merely extensions of existing everyday practices.

The highly theorised and studied subjects of scientific observation and experiment will be considered in the next chapters, but here, examples of different genres of collecting will be examined to illuminate the way this practice operated and – where possible – to situate collecting within domestic space. The examples of collectors are largely drawn from the wealthy classes, who had the greatest scope to acquire, store and display collections. Nonetheless, lower-status individuals took an active role, especially in the realm of natural history collecting, illustrating and publishing.

Specimens

During the eighteenth century, specimen collecting, documenting and illustrating was a widespread activity. As mentioned in Chapter

2, many people kept nature diaries and the collection of information about a locality was considered an entirely appropriate form of empirical practice. The English naturalist and illustrator, James Bolton (1735–99), who will be discussed in greater detail below, published a three-volume text, *An history of fungusses growing about Halifax* (1788–90), and this geographically localised enquiry was typical of many natural history publications of this period.[20] These studies had the obvious advantage of allowing the author to study their own neighbourhood and achieve a comprehensive exposition of its plants or animals. A deep and long-term familiarity generated the detailed knowledge needed for such an undertaking. However, some naturalists did embark on projects of a larger geographical scale, at which point the acquisition of information and specimens from a dispersed network of contacts became crucial. James Sowerby (1757–1822) was one such naturalist and illustrator who created 2,592 hand-coloured engravings for the ambitious thirty-six-volume *English botany*, which was published at the turn of the nineteenth century over a period of twenty-three years.[21] Sowerby's career had begun with an education at the Royal Academy and a specialisation in flower painting, but he later pursued interests in natural history and mineralogy. Surviving correspondence between Sowerby and the Irish naturalist, John Templeton (1766–1825), reveals the tactics and networks of specimen acquisition that fuelled natural historical research in this era.

The Sowerby–Templeton correspondence evinces a number of common characteristics of specimen collecting. Like other examples in Chapter 2, Templeton was an avid journal-keeper as well as a letter-writer, and he regularly included details on the weather in his journal alongside his main subject. With a wealthy mercantile background, Templeton maintained connections with a range of English naturalists – including Joseph Banks – via correspondence and was an important supporter of the establishment of the Belfast Botanic Gardens. Whilst Sowerby was the skilled artist of the two, Templeton included many sketches of plants in his own journal and he stitched into his 1806 volume a pamphlet by the painter, Edward Dayes, entitled 'Essay on the usefulness of drawing', which had been printed in the local press.[22] The essay argued that drawing provided a foundation for painting, but – more than that – it offered a range of advantages to everyday life: 'drawing opens the mind ... it teaches to think'.[23] Templeton's letters and notebooks offer a combination of visual and textual accounts of the subjects of his enquiry and make the point that record-keeping, especially concerning collections, could take a pictorial form.

The correspondence between these two men reveals many of the dynamics of naturalist networks of this period. Publication projects relied on the wider access to observations and specimens offered by friends and colleagues across the country and, in some cases, much further afield. On 26 July 1798, Templeton congratulated Sowerby on his 'scheme of having a true British Museum' of natural history and hoped that 'by sending you still some of the rarest [specimens] that some may be of use either to embellish your publication or increase your Collection'.[24] Templeton was not Sowerby's only contact in Ireland and he encouraged his English friend to exploit other connections on the island to secure what was necessary:

> I will endeavour to procure every thing for you which is the Natural product of this Country, I have seen some specimens of the Irish Gold [*Thuja plicata*, western red cedar] and if you cannot get it by a friend in Dublin I would send there for a piece, but as I may not for a long time have an opportunity of chusing myself a proper specimen ... I wish you to take advantage of any friend upon the spot to get that article for you.[25]

Templeton further urged Sowerby to 'look at some [specimens] I sent Dr Shaw' at the British Museum, on account of their being 'the Most Curious ones of this part of Ireland', adding, 'I thought to receive an account of them from him [Shaw] before this, for I was very anxious to know the proper Names of some of them, as also whether I was right in my Conjectures about the fish which I sent the specimens of to him'.[26] Thus, naturalists like Templeton sent specimens to members of their network for a range of reasons – to participate in scholarly exchange, to provide an example that would fill a gap in a colleague's knowledge or collection, to prompt reciprocation and to secure further details about the specimen from an expert. Specimen collecting represented one part of a larger naturalist practice, but the opportunities and imperatives provided by research practised close to home fuelled the development of these networks of exchange.

A letter written by Templeton to Sowerby on 5 July 1797 reveals the difficulties encountered by collectors, including the task of keeping specimens in good condition and the pressure to exchange both specimens and useful information: 'I will delay sending it now in hopes of getting a fresh specimen. I now enclose you *Rhodiola rosea Empetrum nigrum* [a perennial flowering plant] with a berry'.[27] The promise is followed by details of observations made in Templeton's own garden as well as on hill walks in the north of Ireland: 'some plants in my garden had female flowers alone this spring nor could I find any Male Flowers on the Wild Mountain plants'.

Templeton's information betrays the particularities of the Irish landscape, noting the presence of the '*Myrica Gale Monœcius* Variety' of flowering plant – commonly known as 'bog-myrtle' – and commenting on its regular presence in the bogs of his neighbouring countryside.[28] Templeton described how the specimen might suit Sowerby's purpose in illustration – 'to figure with the *Dioecius* Males and a branch with leaves' – and noted that 'by the Specimen you will see it varies greatly in the form of its leaves, in the broad leaved Variety the Rose Color is much more brilliant than the others they are both found on Bogs in the County of Down Near Donaghadee & Grey abbey'.[29] The exchange speaks to Templeton's familiarity with Sowerby's art practice and his efforts to meet the needs of his colleague.

It seems that Sowerby was also conscious of the pressure he placed upon his correspondents. On 25 July 1810, he apologised for giving 'trouble to my friends' in his persistent collecting of specimens but stressed that he 'had rather be obliged to Gentlemen for some things, especially if I can make a return, than go to a dealer where there is no obligation'.[30] Thus, semi-social relationships which engendered reciprocity were, in this case, preferred to straightforward transactions. The mutuality of shared endeavour was a strong motivation for these men. In January 1820, Templeton's letter arrived with Sowerby via the hands of 'perhaps the best Conchologist in this Country [Ireland]'. Templeton noted that this man 'wished to present you with what he think a distinct species of Trochus', but encouraged his reader further with the promise of other specimens and the benefit of yet another useful contact: 'I believe he had however some other truly distinct species and as he is a diligent Collector once you are acquainted I have hoped he will not be a useless acquaintance to a Man of Science.'[31] In these letters, frequent and specific demands are made, especially of Templeton's time and resources, however, this exchange of specimens and knowledge was skilfully eased by a careful combination of apology, deference and flattery.

Although geographically disparate, Templeton and Sowerby occupied a similar social stratum. However, specimen collecting often produced interactions between individuals with divergent class identities. For example, famously, Margaret Cavendish Bentinck, second Duchess of Portland (1715–85) amassed a vast collection, and the preface to the sale catalogue for the 1786 auction of that collection claimed 'to have had *every unknown* Species described and published to the World'.[32] Portland's 'totalizing interests of curiosity' were able to incorporate highly disparate categories of objects in one whole.[33] However, the acquisition of this remarkable collection embraced a diverse range of intermediaries. Whilst expensive and imported ceramics formed a core part of the Duchess's project, sourcing natural specimens entailed a network of helpful contacts. James Bolton was one such contact, a naturalist, botanist, mycologist and illustrator. The younger son of a weaver, it is not known if Bolton received any formal education, and he was a self-taught artist. Having first worked in the same trade as his father, Bolton later became an art teacher and, in the last years of his life, a publican in Luddenden Foot in Calderdale, West Yorkshire.[34] As a naturalist, Bolton benefitted from the patronage of Portland and, in a letter most likely written around 1780, it is clear that he facilitated her interests in birds. Bolton would later publish a two-volume work on British songbirds, *Harmonia ruralis*.[35]

Bolton's correspondence with Portland is packed with the details of his observations of a number of different species of birdlife. His letter opens with the boast that 'My success in regard to birds has prompted me to Write to your Grace and to Hope for Pardon.' The letter was, in this sense, part bid for patronage and part exercise in natural historical knowledge-sharing. Bolton included two of his own drawings of buntings with the letter and offered to provide more, impressing upon the Duchess his skills as an observer and illustrator. He also intended to send her 'a fine pair' of crop bills he had shot, which he hoped would 'afford an agreeable pleasure to your Grace, They being so very different in Colour that one could scarce believe them the same were it not for the Bill & Feet'.[36] However, this was not the first communication between the naturalist and the collector; Bolton referred to 'your Graces command concerning Insects' and hoped to be able to 'send up a few [specimens] about the end of summer which are not found in Cabinets'.[37] Bolton's letter indicated that he could secure a range of specimens in the future, including 'a fine cock bird of the Tawney Bunting', which he described as 'a beautiful and lovely Bird and never before seen by me'. He also had in his possession a mountain bunting, a greater middle and lesser spotted woodpecker and a 'Coalmouse' meaning a coal tit.[38] Accompanying each offer of a specimen were details about the birds that Bolton found notable, for example, the woodpeckers he regarded as 'beautifull & rare' but also 'imperfectly described by the writers'. Meanwhile, the coalmouse he declared 'absolutely Different from the marsh titmouse with which it has been confounded'.[39] In this way, Bolton stressed the importance of first-hand examination of such specimens and, if not, the use of high-quality illustrations of the kind he could offer.

Bolton's letter is a piece of advocacy for the study of birds, but especially those of 'the smaller kinds' and he bemoaned the 'great imperfection and deficiency' in the understanding of these creatures and the persistent 'errors in respect to colour, which cannot be corrected any other way than by writing new descriptions immediately from life'. Bolton insisted that there 'are many new Observations to be made regarding their manners, haunts, food, Nests, Eggs, Young, times place &c. &c.' but reassured the Duchess that 'from long observation & strict inquiry' he was 'pretty well acquainted' with these details.[40] Although Bolton had himself secured a 'fine specimen' of a mountain bunting, he also noted that this was a bird that 'Mr Pennant' had never seen, referring to the well-known Welsh naturalist Thomas Pennant (1726–98). Thus, he nodded to his own connections within naturalist networks but also to his privileged access to unusual specimens and promised to 'omit no opportunity of procuring such nests and eggs as are rare or beautifull'.[41] By using a combination of convincing ornithological detail based on first-hand observation, alongside offers of drawings and rare and beautiful specimens, Bolton endeavoured to secure Portland's interest and, ultimately, her financial support. Unfortunately, the Duchess's death in 1785 put an end to this arrangement and Bolton would only publish his work on birds a decade later.

This example illuminates the socially encompassing character of natural history in this period. James Bolton was brought up on a farm near the Calder Valley and around three miles from the hub of textile industries, Halifax. However, the Bolton family produced two naturalists, as James's older brother, Thomas (1722–78), was also active in this field, focusing his energies on entomology and ornithology. It was Thomas Bolton's name that was given to a large species of dragonfly found in Britain, *Cordulegaster boltonii*, in recognition of his having collected the specimen. The fact that two brothers from obscure origins were both involved in natural history activities has led to some confusion about the attribution of works, and it also seems likely that they collaborated on some projects.[42]

When James Bolton married in 1768 and started a family, he fostered this self-taught occupation in his children, and his sons certainly participated in natural history observing and collecting.[43] Similarly, James Sowerby – discussed above – involved his children in natural history, leading to his son James de Carle Sowerby (1787–81), and subsequent generations, making a contribution in this field. Like other trades and occupations, the home acted not only to accommodate the pursuit but also as a place of transmission. Children growing up in the household of a naturalist or collector easily acquired those skills and knowledge which was – in turn – underpinned by the many other complementary aptitudes developed at home.

In these examples, it becomes clear that the observation and documentation of natural history was an accessible scientific pursuit in a number of ways. Not only was a focus on a local area entirely justified, even preferred, but the acquisition of skills and knowledge was not entirely dependent on an expensive education or a family with financial resources. Whilst the Bolton brothers could be viewed as exceptions, their life histories and accomplishments speak to what was possible from a modest home a few miles distant from even a market town. The networks of contact between naturalists in this period were inclusive precisely because of the need to secure geographically dispersed and rare examples of plants and animals. Moreover, naturalism was a calling that was passed from one generation to the next, a process eased by a domestic environment that both accommodated and demanded the development of skill and knowledge. The larger project to fully 'know' nature was impossible without high levels of participation and verbal and written exchange across these islands in this period.

Instruments

Another example of a collection amassed in a wealthy household is that of Margaret, Lady Clive (née Maskelyne, 1735–1817), although the focus in this case was instruments of astronomy. The Maskelyne family's close connections with the East India Company led to Margaret's marriage to the military leader Robert Clive (1725–75) in Madras in 1753. Despite losing her mother, Elizabeth Maskelyne, at the age of thirteen, Lady Clive's adult letters recalled the importance of this relationship in initiating her own interests in poetry and astronomy. The maternal influence in the Maskelyne household was also evident in the path taken by Clive's brother, Nevil Maskelyne, who became Astronomer Royal and a member of the Board of Longitude and who published an annual nautical almanac with tables that facilitated the lunar-distance method of finding longitude at sea. As a child, Clive had been tasked with copying out her mother's poems and, in adult life, she often quoted them in her letters.[44] The Clives' fortunes changed in the late 1760s when Robert Clive became embroiled in a political scandal and his actions in India became subject to a public enquiry in 1772–73. Clive was forced to defend not only his reputation but also his private fortune. Although Clive won this battle, he died suddenly in 1774, leaving Lady Clive to lead her remaining decades at Oakly Park in Shropshire.[45]

Margaret Clive's letters dating from this later period of her life 1775–1805 survive in the British Library. The extant letters are addressed to her brothers Edmund (d. 1775) and Nevil Maskelyne (1732–1811), her sister-in-law Sophia (Nevil's wife, 1752–1821) and her niece Margaret (Nevil and Sophia's daughter, 1785–1858). Her correspondence documents her engagement with astronomy and her collection of globes and telescopes at Oakly Park. Astronomy became increasingly accessible to women in the later eighteenth and nineteenth centuries, with educators such as Benjamin Martin (1705–82) and James Ferguson (1710–76), giving public lectures and offering courses to diverse and often female audiences. Clive's access to astronomy was of course furthered by her family relationships and her incredible wealth. Moreover, the connection between astronomy and poetry revealed by Clive's correspondence was a common one in this period, with Ferguson's pupil Anne Lofft (née

Emlyn, 1753–1801) receiving poems from her husband on the theme of her love of astronomy.[46]

Clive's letters of 1806 to 1808 refer regularly to her growing collection of terrestrial and celestial globes. She described how she used and valued them as instruments and personal possessions – possessions she would sometimes give away as significant gifts to favoured relatives. She was regularly looking for new acquisitions for her collection and would send globes to a Mr Dolland for repair and refurbishment. As a result, letters to her niece – Margaret Maskelyne, daughter of Nevil Maskelyne – contain regular requests and updates concerning the travel of her treasured globes to and from the repair shop and the possibility of buying additional pieces for her collection. In a note squeezed into the margin of an 1806 letter, Clive asked her niece to find out if her father had 'seen a pair of Globes fit [for] my use?'. She simultaneously bemoaned the loss of the painted imagery from her existing German-made globes, saying 'the Bears alas! alas! now almost bare after this scrubbing'.[47] When they had been with Mr Dolland for some time, Clive decided to 'hurry Mr Dolland' to send back her 't[w]o lamented old friends', resolving to 'love them as before, & not mind the shabby dresses of the poor Bears, who could ill bear the scrubbing of soap & water; & who had survived three or four ugly falls'.[48] This and other globes in Clive's collection would enter Dolland's workshop on a regular basis, and once there she would anxiously await their return.[49]

On 2 March 1806, Clive wrote to Margaret Maskelyne saying, 'I had hoped every mail & waggon day would have gratified my longing eyes with the sight of the globe'.[50] Clearly, Dolland took time over his work because on Easter Day that same year, Clive revealed that her 'old globe' was still in his hands and she worried that 'none of the Wise Artificers can do any thing to it'. Nonetheless, the traffic in globes continued apace, as Clive rejoiced 'to hear my fair terrestrial Globe is coming' and felt 'tenderly about my Globe now on the road to Mr Dolland' insisting that she 'must [have] it back, or have one to follow with it, covered with any map Doctor Maskelyne will conjure up for me'.[51] Mr Dolland's work extended beyond the technical task of painting and re-painting maps onto the globes, he was also directed to 'brighten the brass parts, & to make them as nicely clean as he can'.[52] These comments reveal a trade not only in new and antique globes, but also the role of artisans in a variety of design and repair work. These objects were valued for their insights into the terrestrial and celestial worlds, but also for their visual appearance within a domestic environment.

Newer additions to Clive's collection included a celestial globe which she dedicated to her brother Nevil Maskelyne and a terrestrial one she dedicated to the naturalist, Sir Joseph Banks, which she reported positioning between two chairs in a principal room of the house and 'upon high stands' so that 'the room will look very handsome'. Clive noted that 'the Cedar tables may remain against the door, as it is so ornamental to the room'.[53] Elsewhere, Clive mentioned rejecting a chintz sofa to put a globe in its place. Pleased with the result, she reported, 'it looks delightfully just facing me as I sit with my back to the fire in my morning working room, & I can draw it about just as I please'.[54] In December 1806, Clive announced, 'Now I am finely set up with Globes and all the honour of my family.'[55] These globes were clearly appreciated for their aesthetic contribution to a room as prized pieces of furniture, as well as for their more explicit function. For Clive, they became household personalities that accompanied her about her domestic business.

Globes formed a central interest for Clive, but her letters reveal her engagement with other astronomical instruments and the relationships that existed between these objects. On 6 April 1805, Clive commented that a 'little quadrant will be a fine play-thing'.[56] In her Easter Day letter of 1806, Clive mentioned 'the Quadrant of Altitude, which belongs to that lovely Globe Dr John Walsh so kindly gave me'.[57] Moreover, 'pocket globes' were also the subject of regular mentions.[58] Clearly the imposing aesthetic contribution of full-sized globes represented part of their appeal, but the attraction of pocket-sized versions suggested that Clive valued the globes as portable objects or that her urge to collect extended to all sizes and categories.

This collection was put to work in service of Clive's scientific interests. She read about the latest discoveries in astronomy – in particular, the new discoveries of stars in the early 1800s. To gain the best chance of sighting a new addition to the known constellation, Clive would move her telescope from place to place in her house. Unfortunately, her instrument was not powerful enough to capture all such discoveries and, in 1807, she was disappointed in her search for an asteroid named Vesta, which Clive referred to as

'Dr Olbers the second'.[59] She likewise tried and failed to see Uranus. However, nearer planets were in her sights. In 1804 she described the conjunction of Venus and Saturn (when viewed from Earth they appear to align in a similar part of the sky) as 'Venus pulling Saturn's ring in a rude manner', demanded that someone draw 'the exact & precise manner of his impertinence' and contributed her own sketch of the phenomena to this purpose, hoping that it might resemble what had been seen by astronomers at Greenwich.

However, Clive's light-hearted style belied a serious engagement with her subject and the letters show that aunt and niece collaborated to gain greater insight into the night sky. On 1 July 1804, Clive admitted to begging her niece 'again and again' to 'relate to me your adventures and observations' in respect of astronomy.[60] Her interest related specifically to the question of viewing the newer planets through a particular telescope 'before Jupiter gets too far from my favorite portion of the heavens'.[61] Clive also referred to an 'assistant' who was helping her to locate the new planets without success, suggesting that a household servant had been directed to this task. On 4 May 1807, Clive requested, 'When you favour me with accounts of the appearances of new planets; let me beg you will specify exactly on what part of the Heavens they shew themselves.'[62] Later the same month, Clive pressed her niece further:

> I must teize [tease] you for an answer to my question, where was seen Olbers's new planet, which I have named Dr Olbers. I beg you, my Margaret, to give me an exact account of this new planet, as likewise of the other 3, whose orbits you say are all between Mars, & Jupiter.[63]

Likewise in early 1808, in a note added after her signature, Clive demanded, 'When you tell me of the Comet's places, say in that part exactly of a constellation, as well as of it's [*sic*] right assension &c. pray tell me if the new Comet will or does appear here, pray do!'[64] These letters imply that both Margarets were viewing the same celestial phenomena at the same time, often using instruments known to both of them and comparing notes in their regular correspondence. However, there was a larger network of individuals who Clive encouraged to join them in these activities.

On 9 December 1804, Margaret Clive was entertaining. At three o'clock in the early hours of the morning she 'caused three families to rise & keep themselves awake, to watch Venus & Saturn, & they saw them plainly'.[65] A week later, Clive reported rising twice in the night to catch a glimpse of these same planets and failing to do so. However, three houseguests 'saw the two planets & a very pretty creature besides', prompting Clive to comment that 'Had I known that the 3 worthy observers would have seen any thing so plainly, my telescope should have been carried over night to the best spot.'[66] Clearly, Clive regularly entertained people who were either engaged with astronomy already or encouraged by Clive's own enthusiasm and the ready availability of a telescope.

Clive and Maskelyne were highly privileged in terms of the money and social connections they could bring to this enterprise. Very few could boast access to the Astronomer Royal and his instruments and knowledge as tools of their enquiry. Nonetheless, these women relied upon cheap printed resources, such as an ephemeris,[67] that made astronomy accessible across a wide social spectrum, as will be explored in the next chapter. In fact, Clive complained of the barrier cost placed on other facets of astronomical information, 'I really think astronomical observations ought not to be charged, but go free', suggesting an interest in broadening participation in this field.[68]

Whilst Margaret Clive registered an interest in the accessibility of her favourite pursuit, her most concerted efforts were put into her own family's ongoing specialism in this field. Significantly, in 1792 Clive left instructions for the 'Maskelyne and Banks' pair of globes to be given to her brother, Nevil, after her death. She added, 'I hope you will give them to your sweet child by & by', referring to her correspondent Margaret Maskelyne.[69] Clive's insistence that this treasured globe should ultimately come into her niece's hands underlined the female tradition of studying astronomy in the Maskelyne family. The globes were gifted to father and daughter alongside 'the pianoforte, the old family cabinet ... together with the antique cat',[70] revealing the categorisation of scientific instruments as prized domestic objects of practical, aesthetic but also sentimental value. Clive's letter to her brother to express these wishes was to stand in place of lines in a will, she admitted, 'I ought, long ago, to have said this but indolence about altering a paragraph in my will ... has occasioned me to say nothing about it till now.' The letter was intended to be preserved and, Clive hoped, accepted as the 'trifling marks of a sister's love'.[71]

Clive's engagement with astronomy incorporated both her literary sensibilities and her commitment to scientific observation of natural phenomena. Her collection of instruments attended to scientific need, domestic aesthetics and the collector-connoisseur's urge to acquire in number as well as quality. Her collection ranged from large globes, set on pedestals, to a three-inch pocket globe. For Clive, astronomy was an explicitly domestic activity, but also one that had been transmitted from one generation to the next and, notably, down the female line.

Antiquities, art, books and manuscripts

In the papers of the Earls of Malmesbury in Hampshire Record Office lie a series of catalogues and inventories belonging to the English MP and grammarian, James Harris (1709–80).[72] Harris authored several works, including *Three treatises – on art; on music, painting and poetry; and on happiness* (1741) and *Hermes, a philosophical inquiry concerning universal grammar* (1751) and was also a composer of music. In 1763, he became an elected fellow of the Royal Society, he was a Trustee of the British Museum, a Lord Commissioner of the Admiralty, a member of the Board of Treasury and comptroller to Queen Charlotte. In support of his wide-ranging interests, Harris became a keen collector of art, manuscripts and antiquarian artefacts and the extant catalogues – three stitched notebooks – contain lists of books, prints, etchings, views and descriptions of places, ruins, antiquities and maps alongside the details of where they were kept within his home at the Close, Salisbury. As his biographer notes, 'Harris has come down to us, so far, like a piece of unsorted intellectual debris'. An influential figure in his own time and an active enabler of other artists and thinkers, Harris's notes on his own collection offer a glimpse of a large domestic collection closely connected with a scholarly life.[73]

The first of the notebooks is entitled a 'Catalogue of books of prints' and includes information about Harris's etchings, maps and depictions of ruins; the second is, similarly, a 'Catalogue of drawings, prints and etchings'. Both are dated on the back cover as having been compiled in 1780.[74] The third notebook contains details of the 'Arrangement of papers, keys &c. in various chests' and was comprised several years earlier, in 1776.[75] These notebooks offer insight into the way collections were organised, housed and secured in the context of an affluent eighteenth-century home.

Harris's Salisbury home had undergone significant remodelling in the early 1700s and the house included an impressive, first-floor library, which he had decorated in the contemporary Gothic style. The contents of the library were described, in 1780, as one of the best private collections in Europe and most of the locations recorded in Harris's notebooks are in this galleried room with a fireplace and nine 'Recesses or Apartments for Books'.[76] Besides the volumes displayed in the, presumably shelved, recesses – there were a series of chests that contained the prints, manuscripts and other artefacts.[77] To orientate the reader, Harris mapped the storage of his collection onto the layout of the library as follows:

> In the Library, beginning from the left hand of the Chimney, & so passing on to the right, are nine Recesses or Apartments for Books. There are also in the Gallery two Chests; the farthest from the Library Door, called Chest the first, being for Bound Books of Prints, Drawings, Antiquities &cr; the other, called Chest the second, being for Drawings, Prints, & Etchings, detached & the greater part in Porto Folio's.[78]

Harris used abbreviations to refer to specific locations, noting that 'In the following Catalogue, R.1. R.2. &cr denote the respective Recesses in the Library's C.1. & C.2 denote the two Chests in the Gallery.'[79] Chests were usually locked, and the third notebook reveals the locations of keys and master keys for boxes, chests and rooms containing parts of the collection. It is perhaps no great surprise that the library was the centre of Harris's collection and a room he had devoted resources and care to remodelling. In keeping with philosophical discourse, Harris compared the mind to a library with a single classificatory system – arguing that the mind should be 'furnished, like a good Library, with proper Cells or Apartments' in which 'our Ideas both of Being and its Attributes' can be filed, ready for recall when required.[80]

Whilst the library was the primary location for Harris's collection, the catalogues also refer to an exchequer room, a chapel and two closets – adjacent to the library and the chapel respectively. In many cases, the locations of artefacts are given in precise detail, such as a 'Cellarii Geographia' being housed in 'R.5 – lower shelf'.[81] Similarly, 'Large Medals of eminent Persons in Russia' made of copper and acquired by Harris's daughter, Gertrude, in October 1779, were to be found 'in the lowest Drawer of the second Chest'.[82] When describing the position of a 'Fine Mezzotinto' on the upper shelf of the second chest, Harris noted that they share a shelf with 'six Impressions in wax, given me by that excellent artist Mr. Wray, from Intaglio's cut by himself'.[83] In the third notebook, the relocation of articles from one place to another was considered worth recording. On 20 August 1777, 'I removed, from the Closet in my Library' one quarto manuscript, five folio manuscripts, a further couple of marked folio manuscripts and 'placed [them] in my Exchequer on the shelf over the door'.[84] Whilst the category of artefact comprises one part of the spatial organisation, security also seems to be a deciding factor. In the exchequer, Harris kept keys to the chapel, the harpsichord, the library and the library closet. However, he kept the keys to the exchequer itself, the chapel closet and a master key for the library and library closet in his pocket, close at hand.[85] Whilst these notebooks were practical guides to accessing and understanding this diverse and valuable collection, they occasionally hint at the connection Harris made between himself and his objects. For example, in the front of the notebook describing the arrangement of papers and keys, Harris pinned a declaration, dated 1776, by Matthew Burgate that he had made groundless aspersions against James Harris.[86] By combining these details, Harris articulated something of the sense of self that was harboured by a personal collection, relating the safekeeping of his reputation to the security of his prized things.

Beyond the accessibility of Harris's collection, the two catalogues occasionally attend to the circumstances of purchase and provenance of individual artefacts or groups of objects. They also highlight the Harris family's elite social connections, the first Catalogue recording that the 'Antiquities of Herculaneum' were acquired via Harris's son as a gift from the King of Spain.[87] Harris notes that the 'first volume contains a Catalogue of the Curiositys found there', 'the four next volumes contain the Pictures; the two last, the statues & bronzes'.[88] Other acquisitions were sourced closer to home: 'In May 1776 I bought at Boydel's fine Printshop in Cheapside the Regulus and the Wolfe, both after the Pictures of West; also two elegant oval Landscapes (the figures graved by Basto-Cozzi) both after the Paintings of Leuterberg. The Regulus, being an original Proof, cost me three Guineas.'[89] Harris was spending part of a legacy of five guineas that he had been left by his 'worthy Friend, Dr John Hoadly, Master of St Cross, & Chancellr of the Diocese of Winchester'.[90] The choice of artwork was not an idle one as Harris concluded that 'these beautifull Works were things, which I knew I should often contemplate' and in so doing 'reflect on the Affability, the Ingenuity, and the Virtue of that good man'.[91]

Harris clearly benefitted from his wide-ranging social contacts to facilitate his collecting. He had, for example, secured 'an Impression in wax from a small Gemm' of 'the pensive Hercules' from Mr Hoare of Bath, who had himself acquired it in Italy. Harris commented on this 'curious' artefact, suggesting that it was a copy of the Ancient Greek sculptor, Lysippos, whose original works had been 'destroyed (together with many more invaluable statues) in the year 1205 by the Barbarians of Baldwyn's Crusade, when they sackt Constantinople'.[92] The catalogue includes a further cross-reference to a work of reference with relevant page numbers. Thus, the catalogues speak not only to the importance of given objects to Harris himself, but also to the network of friends with which Harris shared his research and interest in classical antiquity. These documents appear to have been designed primarily as an aid to memory for the organisation of the collection as a whole, but the references to the particulars of prized items, the circumstances of purchase or the personal associations of given artefacts suggest that Harris's social and intellectual networks intruded on his task of cataloguing. It is possible that he had an eye towards the use of his catalogues by those tasked with distributing his personal effects at death. However, the ad hoc quality of the inclusion of these details implies something less planned and more intuitive. They provide a glimpse of the networks of association that surrounded his collection and stretched far beyond the recesses of his library.

In these examples of collectors and collecting, several themes emerge. The importance of correspondence networks and the acquisition, via intermediaries, of rare and special specimens and artefacts have been well-studied. For the wealthy, foreign travel also offered opportunities to collect, especially works of art and historical artefacts. However, here it is clear that naturalists could turn a life rooted in a particular location to their analytical advantage. The inability to travel was certainly no barrier to this kind of enquiry. As Chapter 5 will show, the descriptive intensity made possible by prolonged engagement with particular species of wildlife was of real benefit. Whilst the collectors discussed here are drawn from the elites, they do offer considerable detail about the domestic arrangement and safeguarding of their objects. The way collections were arranged in their homes was highly meaningful, evoking important aspects of the artefacts' histories, aesthetic qualities and social resonance. The curiosity, alongside the specimens and artefacts, was something to be passed on to future generations – generating linkages across time as well as space.

Conclusion

Collecting, especially on the scale of elite collectors in this period, is classed as an activity that transcends the parameters of home and occupies the grander spaces of public, intellectual life. However, for those who did amass and did so at some scale, the home was clearly *the* location for this pursuit. The household was not a passive container in which objects of meaning could be placed. On the contrary, the home offered collectors the possibility of displaying, organising, safe-keeping and interpreting their possessions and doing so iteratively over time. By identifying the role of the home in practices of foundational importance to many forms of enquiry in this period, their meanings are transformed. Instead of artificially separating the 'scientific' from the 'everyday', the discussion here has shown that they formed part of a piece and that the former can only be fully understood by acknowledging its domestic habits and ethos. The next chapter will move on to consider the activity of observation and will do so by attending to the ways it was shaped by the characteristics and contingencies of home.

Notes

1 James Delbourgo, *Collecting the world: The life and curiosity of Hans Sloane* (London: Penguin Books, 2017); Jennifer Potter, *Strange blooms:*

The curious lives and adventures of the John Tradescants (London: Atlantic Books, 2014).

2 Sean Silver, *The mind is a collection: Case studies in eighteenth-century thought* (Philadelphia, PA: University of Pennsylvania Press, 2015), p. 2.

3 Crystal B. Lake, *Artifacts: How we think and write about found objects* (Baltimore, MD: Johns Hopkins University Press, 2020), p. 66. However, Gillian Williamson emphasises the military connotations of this word (relating to the storehouse as arsenal or armoury) and further argues that seventeenth-century titles that used this term typically addressed the needs of tradesmen: *British masculinity in the* Gentleman's Magazine, *1731–1815* (Basingstoke: Palgrave Macmillan, 2016), p. 44.

4 R. G. W. Anderson, M. L. Caygill, A. G. MacGregor and L. Syson (eds), *Enlightening the British: Knowledge, discovery and the museum in the eighteenth century* (London: British Museum Press, 2003); Lake, *Artifacts*, pp. 47–52.

5 For further details about this location, see www.british-history.ac.uk/survey-london/vol2/pt1/pp61-64 (accessed 11 November 2021).

6 Lake, *Artifacts*, pp. 47–63.

7 Clare Haynes, 'A "natural" exhibitioner: Sir Ashton Lever and his *Holosphsikon*', *British Journal for Eighteenth-Century Studies*, 24 (2001), pp. 1–14.

8 Elizabeth Baigent, 'Bullock, William (bap. 1773, d. 1849) naturalist and antiquary', *Oxford dictionary of national biography online* (Oxford: Oxford University Press, 2004): https://doi.org/10.1093/ref:odnb/3923

9 Alice Marples, 'Medical practitioners as collectors and communicators of natural history in Ireland, 1680–1750' in John Cunningham (ed.), *Early modern Ireland and the world of medicine: Practitioners, collectors and contexts* (Manchester: Manchester University Press, 2019), pp. 147–64; for civic and scholarly culture in Ireland, see also Kelly and Powell, *Clubs and societies*.

10 *Old Bailey proceedings online* (www.oldbaileyonline.org, version 8.0, 11 November 2021), January 1829, trial of Benjamin Williams, John Brinkley (t18290115-47); Watson was a stationer with a shop on the High Street, Wapping in London.

11 *Old Bailey proceedings online* (www.oldbaileyonline.org, version 8.0, 11 November 2021), January 1829, trial of Benjamin Williams, John Brinkley (t18290115-47); the accused were found not guilty.

12 *Old Bailey proceedings online* (www.oldbaileyonline.org, version 8.0, 11 November 2021), January 1829, trial of Benjamin Williams, John Brinkley (t18290115-47).

13 In the eighteenth century, Hemmings Row was positioned between Castle Street and St Martin's Lane, but the area was reconfigured in 1886 with the building of Charing Cross – at this point Hemmings Row became part of St Martin's Place; since the 1660s there had been a workhouse at Hemmings Row and a new and larger workhouse was built on this same site in 1772. See G. H. Gater and F. R. Hiorns (eds), 'Hemmings Row and Castle Street' in *Survey of London: Vol. 20, St Martin-in-the-Fields, Pt III: Trafalgar Square and neighbourhood* (London: London County Council, 1940), pp. 112–114: *British history online*, www.british-history.ac.uk/survey-london/vol20/pt3/pp112-114 (accessed 11 November 2021).

14 *Old Bailey proceedings online* (www.oldbaileyonline.org, version 8.0, 11 November 2021), October 1784, trial of Robert Artz, Thomas Gore (t17841020-9).

15 *Ibid.*; in this case the accused were found guilty and sentenced to death.

16 Alice Marples, 'James Petiver's "joynt-stock": Middling agency in urban collecting networks', *Notes and Records: The Royal Society Journal of the History of Science*, 74:2 (2020), pp. 239–58; Marples, 'Medical practitioners'.

17 Delbourgo, *Collecting the world*, pp. 202–11; see also Michael A. Salmon, Peter Marren and Basil Harley, *The aurelian legacy: British butterflies and their collectors* (Berkeley and Los Angeles, CA: University of California Press, 2000), pp. 106–8; Richard Coulton, '"What he hath gather'd together shall not be lost": Remembering James Petiver', *Notes and Records*, 74 (2020), pp. 189–211; Tim Somers, *Ephemeral print culture in early modern England: Sociability, politics and collecting* (Martlesham: The Boydell Press, 2021).

18 Benedict, 'Collecting trouble', p. 118.

19 Sloboda, 'Displaying materials', p. 459.

20 M. Seaward, 'Bolton, James (*bap.* 1735, *d.* 1799)', *Oxford dictionary of national biography online* (Oxford: Oxford University Press, 2004): https://doi.org/10.1093/ref:odnb/2803; a supplement to this work was published in 1791.

21 This publication project was a collaboration with the botanist, Sir James Edward Smith, who provided the text. Sowerby's son, James de Carle Sowerby (1787–1871), continued the project after his father's death.

22 Ulster Museum (hereafter UM), Templeton MSS, S54.

23 UM, Templeton MSS, S54: Edward Dayes, 'Essay on the usefulness of drawing', *Belfast News-Letter* (19 Jan. 1809), p. 9.

24 UM, MS70: John Templeton to James Sowerby, 26 Jul. 1798.

25 *Ibid.*

26 *Ibid.*
27 UM, MS70: John Templeton to James Sowerby, 5 Jul. 1797.
28 *Ibid.*; Templeton lived at Orange Grove in Belfast with ready access to the hills surrounding the city and the coastlines of Antrim, Down and the Ards peninsula.
29 UM, MS70: John Templeton to James Sowerby, 5 Jul. 1797.
30 UM, MS70: James Sowerby to John Templeton, 25 Jul. 1810.
31 UM, MS70: John Templeton to James Sowerby, 16 Jan. 1820.
32 *A catalogue of the Portland museum* (London, 1786), p. iii, quoted in Sloboda, 'Displaying materials', p. 459.
33 Sloboda, 'Displaying materials', p. 460.
34 Seaward, 'Bolton, James'; see also John Edmondson, 'New insights into John Bolton of Halifax', *Mycologist*, 9:4 (1995), pp. 174–8.
35 James Bolton, *Harmonia ruralis; or, an essay towards a natural history of British song birds*, 2 vols. (Halifax, 1794–96).
36 University of Nottingham Special Collections, PWE5: James Bolton to Duchess of Portland, n.d.
37 *Ibid.*
38 *Ibid.*
39 *Ibid.*
40 *Ibid.*
41 *Ibid.*
42 Seaward, 'Bolton, James'.
43 Edmondson, 'New insights', p. 175.
44 BL, Clive Papers, 'Lady Clive I 1775–1805', Mss Eur Photo Eur 287, fo. 4.
45 H. V. Bowen, 'Clive, Margaret, Lady Clive of Plassey (1735–1817)', *Oxford dictionary of national biography online* (Oxford: Oxford University Press, 2004): https://doi.org/10.1093/ref:odnb/63502
46 Mary T. Brück, *Women in early British and Irish astronomy: Stars and satellites* (London: Springer, 2009), pp. 11–14.
47 BL, Clive Papers, Mss Eur Photo Eur 287, vol. 2, pt. 1: Margaret Clive to Margaret Maskelyne, 18 Jan. 1806.
48 BL, Clive Papers, Mss Eur Photo Eur 287, vol. 2, pt. 1: Margaret Clive to Margaret Maskelyne, n.d. Jun. 1806.
49 See, for example, BL, Clive Papers, Mss Eur Photo Eur 287, vol. 2, pt. 1: Margaret Clive to Margaret Maskelyne, 10 Apr. 1806 (Easter Day).
50 BL, Clive Papers, Mss Eur Photo Eur 287, vol. 2, pt. 1: Margaret Clive to Margaret Maskelyne, 2 Mar. 1806.
51 BL, Clive Papers, Mss Eur Photo Eur 287, vol. 2, pt. 1: Margaret Clive to Margaret Maskelyne, 10 Apr. 1806 (Easter Day).
52 BL, Clive Papers, Mss Eur Photo Eur 287, vol. 2, pt. 1: Margaret Clive to Margaret Maskelyne, n.d. Jun. 1806.

53 *Ibid.*
54 BL, Clive Papers, Mss Eur Photo Eur 287, vol. 2, pt. 1: Margaret Clive to Margaret Maskelyne, 30 Jul. 1806. Elite furniture design changed in this period, offering lighter weight and more mobile furniture that could be easily reconfigured in a room; see Sundberg Wall, *Prose of things*, p. 185.
55 BL, Clive Papers, Mss Eur Photo Eur 287, vol. 2, pt. 1: Margaret Clive to Margaret Maskelyne, 18 Dec. 1806.
56 BL, Clive Papers, Mss Eur Photo Eur 287, vol. 1, pt. 4: Margaret Clive to Margaret Maskelyne, 6 Apr. 1805.
57 BL, Clive Papers, Mss Eur Photo Eur 287, vol. 2, pt. 1: Margaret Clive to Margaret Maskelyne, 10 April 1806 (Easter Day).
58 See, for example, BL, Clive Papers, Mss Eur Photo Eur 287, vol. 1, pt. 3: Margaret Clive to Margaret Maskelyne, 12 Jul. 1806; vol 2, pt. 1: same to same, 19 Jul. 1806.
59 BL, Clive Papers, Mss Eur Photo Eur 287, vol. 2, pt. 2: Margaret Clive to Margaret Maskelyne, 23 May 1807; Heinrich Wilhelm Matthias Olbers (1758–1840) was a German astronomer who discovered the asteroids Pallas and Vesta. At this time asteroids were thought to be planets.
60 BL, Clive Papers, MSS Eur Photo Eur 287, vol. 1, pt. 3: Margaret Clive to Margaret Maskelyne, 1 Jul. 1804.
61 *Ibid.*
62 BL, Clive Papers, Mss Eur Photo Eur 287, vol. 2, pt. 2: Margaret Clive to Margaret Maskelyne, 4 May 1807.
63 BL, Clive Papers, Mss Eur Photo Eur 287, vol. 2, pt. 2: Margaret Clive to Margaret Maskelyne, 18 May 1807; five days later, Clive again pressed, 'I wish to know more particularly, in what sign, & in what degree, you saw Dr Olbers the 2d, for that is the Name I give him'; vol. 2, pt. 2: same to same, 23 May 1807.
64 BL, Clive Papers, MSS Eur Photo Eur 287, vol. 2, pt. 2: Margaret Clive to Margaret Maskelyne, 18 Jan. 1808.
65 BL, Clive Papers, Mss Eur Photo Eur 287, vol. 1, pt. 3: Margaret Clive to Margaret Maskelyne, 9 Dec. 1804.
66 BL, Clive Papers, Mss Eur Photo Eur 287, vol. 1, pt. 3: Margaret Clive to Margaret Maskelyne, 15 Dec. 1804.
67 Use of an ephemeris is mentioned in letters written on 1 Jul. 1804, 12 Jul. 1806 and 4 May 1807.
68 BL, Clive Papers, MSS Eur Photo Eur 287, vol. 2, pt. 1: Margaret Clive to Margaret Maskelyne, 18 Jan. 1806.
69 BL, Clive Papers, Mss Eur Photo Eur 287, vol.1, pt. 1: Margaret Clive to Nevil Maskelyne, 7 Mar. 1792; Clive had named these globes after

great men of science, her own brother – the astronomer royal and the naturalist Joseph Banks (1743–1820).

70 BL, Clive Papers, MSS Eur Photo Eur 287, vol.1, pt. 1: Margaret Clive to Nevil Maskelyne, 7 Mar. 1792.

71 *Ibid.*

72 Hampshire Record Office (hereafter HRO), 9M73/G847–849; Harris was MP for Christchurch, Hampshire (1761–80), see C. T. Probyn, *The sociable humanist: The life and works of James Harris, 1709–1780* (Oxford: Clarendon Press, 1991), p. 1.

73 Probyn, *Sociable humanist*, pp. 2–5; Harris has not enjoyed the lasting reputation his life and works might have warranted, in part because of the nature of his intellectual work but also on account of negative remarks on his character by the eighteenth-century writer and critic *par excellence*, Samuel Johnson.

74 HRO, 9M73/G847–8.

75 HRO, 9M73/G849.

76 The *Gentleman's Magazine*, 51 (1781), p. 24 as quoted in Probyn, *Sociable humanist*, p. 23; HRO, 9M73/G847, fo. 1.

77 Harris employed research assistants in the 1750s and 1760s to make transcriptions of Bodleian, British Museum and Corpus Christi College classical manuscripts; see Probyn, *Sociable humanist*, p. 71.

78 HRO, 9M73/G847, fo. 1.

79 *Ibid.*

80 James Harris, *Philosophical arrangements* (London, 1775), pp. 543–4 as quoted in Probyn, *Sociable humanist*, p. 243; see also Silver, *Mind is a collection.*

81 HRO, 9M73/G847, fo. 1.

82 HRO, 9M73/G848, n.fo. (second page of notebook).

83 *Ibid.*, fo. 1.

84 HRO, 9M73/G849, fo. 2v.

85 *Ibid.*, fo. 15.

86 HRO, 9M73/G849.

87 HRO, 9M73/G847, n.p. (last page).

88 *Ibid.*

89 HRO, 9M73/G848, n.p. (first page).

90 *Ibid.*

91 *Ibid.*

92 *Ibid.*, fo. 1v.

4

Observing

Observation and experiment were central to the practice of eighteenth-century scientific enquiry and both activities relied on sensory experience as a means of acquiring knowledge and as a form of knowledge in itself.[1] Of course, much of this period's investigative ethos relied on intellectual and practice-based developments made in the previous century, under the banner of the 'new science' and, by the 1700s, experience had become a primary method of examination. Whereas medieval thinkers saw personal experience as a way of observing how the world was, for eighteenth-century enquiry – it was a method of analysis. Or, as Francis Bacon had put it, of 'vexing' nature into revealing her secrets.[2] As such, scientific observation has a very long history and can be considered as the sensory technology capable of converting experience into knowledge: 'the most pervasive and fundamental practice of all the modern sciences'.[3] For the people explored in this book, the ways in which experience developed through domestic practice was converted into evidence require further examination.

Despite their joint reliance on experience, observation and experiment were defined in 'contradistinction to one another' by nineteenth- and twentieth-century scientists and thinkers. This led to an under-appreciation of observation in histories of early modern science.[4] Moreover, experiment has been conceptualised as active and embodied in contrast with passive or uncritical observation. This chapter discusses examples of observation and the following chapter will consider experimental activities. However, these practices often acted as two sides of the same coin and were strongly connected within print culture, with publication titles often including both.[5] Many of the individuals examined here were engaged in both observation and experiment. Here, a central case study is used to examine observation as it took place in the homes and lives of two Dublin apprentices who observed the skies in their pursuit of astronomical skill and knowledge.

Paying attention

As Lorraine Daston has emphasised, for eighteenth-century naturalists, 'observation was first and foremost an exercise of attention'.[6] By way of preface to this chapter's main case study, it is worth briefly considering a more famous contemporary example. In the scholarship on natural historical observation, the Hampshire-born and based cleric and diarist, Gilbert White (1720–93), is a major figure. In part, this prominence is due to the role played by White's research in laying the foundations for nineteenth-century naturalists, notably Charles Darwin, to make their famous interventions. However, the longevity of his reputation is also attributable to the 'clarity with which he showed the importance of appreciating not only the kind of attention an object invites but also the manner most appropriate for expressing what is observed'.[7] Like other naturalists of this era, White subscribed to a model of information gathering that was intensely local, best articulated in his major work, *Natural history and antiquities of Selbourne* (1789), seeing himself as one part of a much larger, collective project of empirical research. As indicated in Chapter 2, the relationship between observation and record-keeping was a close one. Regular observers of natural phenomena might take their notes in pen and ink, but such observations also formed a core part of the published compilations of learned societies. As the next two chapters of this book show, contributors of such observations to both societies and a vibrant periodical press were diverse and drawn broadly from society at large.

White conformed to one of several common models of the 'man of science'. An Oxford-educated member of the lower gentry, his father had been a barrister and his grandfather the vicar of Selbourne. Whilst White did not enjoy the freedom over his time of the wealthier landed gentry and aristocracy, his comfortable existence and educational and professional connections smoothed his path to enquiry. White's personal delight in observing nature is present in his text and readers, over many centuries, have been drawn to the immediacy of his descriptions. During his lifetime, White took on a range of curacies in Hampshire and Wiltshire, alongside the offices of Junior Proctor and Dean at the University of Oxford. At the end of his life, White had assumed both his grandfather's role and home as the curate of Selbourne living at The Wakes vicarage.

For the purposes of this chapter, there are several features of White's scientific work that are interesting. Firstly, his descriptive writing not only taught others how to observe nature, but also expressed the excitement that could be generated by this process – even when applied to the familiar contours of a local neighbourhood. His work is an articulation of how careful record-keeping based on personal observation could offer a route to distinguishing truth from superstition.[8] However, in this search for observable fact, White did not discard the valuable evidence of traditional belief and anecdote. In a section on the decline of certain types of game and, in particular, the red deer, White relied on the testimony of 'an old keeper, now alive, named Adams, whose great grandfather ... grandfather, father and self, enjoyed the head keepership of Wolmer forest in succession for more than an hundred years'.[9] Similarly, his observations on the presence of bog oak in the south of England rested on the assurances of 'Old people' who 'have discovered these trees, in the bogs, by the hoar frost, which lay longer over the space where they were concealed, than on the surrounding morass'.[10]

The centrality of a range of practices associated with scientific observation (live sightings of natural phenomena, detailed record-keeping and wider information gathering and exchange) combined with engaging prose, ensured a readership for White's work. However, behind his famous *Natural history* sat a range of journal-keeping and correspondence (the latter reproduced or re-imagined as the basis of that publication). As mentioned in Chapter 2, a version of White's weather diary was published in the *Gentleman's Magazine*, thereby dramatically increasing the reach of his writing and feeding into a print culture that was familiar and accessible to very many eighteenth-century people.

The second dimension of White's work that resonates with the findings of this book is the evocation of sensory experience in his descriptions of wildlife. White's approach differed from natural histories that prioritised sight as a primary sense 'in taxonomic descriptions', making clear instead the importance of a range of senses in observation, his narrative including 'vivid accounts of rancid-smelling bats, stinking snakes, the putrid stench of death, and the sulphurous smell of a blue mist that heralds a thunderstorm'. Likewise, he observed the importance of song in identifying species of bird and the way certain geographical features generate echoes.[11] This attendance to the sensory, both the observed senses of fauna and also the sensory perception of the observer, is a feature of the domestic silkworm breeding discussed in detail in the next chapter.

Finally, despite White's privilege in terms of his access to the resources and networks of institutionalised intellectual life, he still emphasised the accessibility of discovery to ordinary people. His characterisation of the everyday as exotic and mysterious established a context and a motivation for enquiry that many less materially fortunate individuals could relate to and replicate. The value of repetition for the observer of nature was obvious in his work, and rhythms of repeated return to the same objects of enquiry were reflected in the characteristics of many domestic tasks.[12] The practices at the heart of White's naturalism may have been commonplace in this period, but his writing did create a vivid and alluring rationale for the active participation of others in this inherently collective project.[13]

Observing the skies

At first glance, astronomy might be judged an inaccessible science on account of its use of expensive instruments and prerequisite mathematical ability. However, as discussed in Chapter 3, astronomy in Britain thrived outside of the formal institutions of learning and science with participation from many 'amateurs'. Whilst England had the Royal Observatory at Greenwich and Ireland boasted three major observatories in Armagh, Birr Castle and Dunsink, institutionalised astronomy was mainly focused on supplying navigational data or developing the mathematical facet of the field.[14] This left the field of practical research wide open to the interested individual. Moreover, astronomy, from its earliest origins, was a field of enquiry integrally connected with observation and also with one of the key instruments of such, the telescope. From the mid-eighteenth century onwards, there was a rapid increase in understanding of the universe, including the identification of new planets, planetary satellites and asteroids. There was an eager public audience engaged with these celestial discoveries, with travelling lecturers offering talks and courses and publishing prolifically on the subject.

The practicalities of engaging with astronomy depended to some extent on the nature of the science itself. As Peter Dear has highlighted, 'there was no formal methodological separation between observational and the calculational parts' of astronomy, and astronomers were in the habit of turning their own observations into predictive tables and models of celestial movements as a prerequisite for sharing their findings with a wider audience.[15] So, on the one hand, the singular nature of astronomical observation – requiring the observer(s) to be in situ, with the correct equipment at a particular time – denied the possibility of easily, publicly demonstrable observations and experiments as were typical in other fields. However, on the other hand, the proliferation of astronomical observations, calculations and predictions in cheap print made this sphere of enquiry surprisingly open to popular participation.

The close relationship between print and astronomical activity was manifold. In particular, the potential for predicting the future made celestial modelling extremely attractive to a variety of people for a range of purposes. In fact, astrologers relied upon astronomical data to make their popular forecasts – whether that was for the weather, the harvest, matters of health or political ferment. Many famous astronomers of the period had a firm interest and belief in the astrological ramifications of their own observations and calculations.[16] Prior to seventeenth-century developments in the scientific method, astrology was considered to be a systematic and 'scientific' endeavour and astrologers made large claims for their science's ability to explain the natural world and its effects.[17] As discussed in Chapter 2, Isaac Butler saw the weather as integrally connected with celestial phenomena and he was personally involved in the compiling of astrological almanacs. Of course, early modern people were very invested in astrology, as the ability to predict natural phenomena was of critical value to anyone who farmed, fished or raised livestock and many more besides. The strong connections between astrology and religious belief, medical treatment and the analysis of society and politics gave this field unparalleled inroads into people's daily lives and worldviews.

Almanacs were the publications that capitalised most effectively on the popularity of astrological reasoning, and in the early seventeenth century, they were likely the largest category of print culture – sold to a truly mass audience.[18] By the mid-eighteenth century, the almanac had passed its peak; however, there remained widespread popular and scholarly belief in astrology and a large readership for this kind of cheap print publication. In the sixteenth and seventeenth centuries, the Irish print trade had been tightly controlled as compared to the British business. However, by the turn of the century, Dublin had begun to emerge from the absolute control of restrictive patents, thereby allowing the book trade to expand. Despite a very large population of Irish speakers in the country in this period, printing was predominantly in English which was the main language of administration and business in Ireland.[19] This chapter reveals the extent to which lower-status astronomers relied on this format to establish and advance their own observations and calculations.

However, the traditional annual almanac was not the only category of publication that helped its readers build their astronomical skills. Mathematical problem-solving was a mainstay of long-running journals. For example, the *Ladies' Diary* or *Woman's Almanack* which ran from 1704 until 1840 had to remove recipes and other forms of content to make room for mathematical challenges. This publication was not only one of the earliest periodicals aimed at a female readership, but it was also the first of its kind to provide a public forum for mathematical exchange.[20] Explicitly instructional texts, such as manuals, offered their more specialised readerships the option to assemble instruments for use in astronomical calculation. In this way, the readers of an earlier generation of instructional text – John Blagrave's *Mathematical jewel* (1585) – could learn by cutting out pre-prepared templates, glueing them onto board and constructing a usable device. This complicates understandings of ways of knowing that primarily rest on sight, text and reading and those that rely on material literacy, physical manipulation and embodied knowledge.[21] However, these different forms of print culture also reveal a wide and diverse audience for mathematics and, with it, the seedbed for 'amateur' astronomers not only to follow but also to participate in the frenzy of celestial discoveries that took place in the later eighteenth century.

Two such enthusiasts were the young Quakers, Robert Jackson (1748–93) and Thomas Chandlee (dates unknown), both apprentices living in Dublin in the late 1760s. The evidence of their endeavours is contained in correspondence held at the Friends Historical Library in Dublin. Only Jackson's letters survive, and at this time he was a twenty-year-old apprentice to his father, the printer and publisher Isaac Jackson of The Globe, Meath Street in Dublin who was also an official printer for the Quakers.[22] The letters were written over the period 1768–69 and document Jackson's role as tutor to Chandlee on matters of astronomy. Chandlee was the son of merchant Thomas Chandlee senior of Athy in County Kildare, but he was apprenticed to Robert Fayle, a linen draper in Bride Street – just a few streets away from Jackson's place of residence and work. Jackson completed his seven-year apprenticeship in March 1769 and began work as a journeyman.[23]

These were a pair of young men, working hard to establish themselves within a trade and living in houses with their master, his family, other apprentices and most probably servants. At the time of Robert Jackson's apprenticeship, his father also oversaw the work of Thomas Byrne (either as an apprentice or a journeyman) and another apprentice who absconded after five years of service.[24] Clearly apprenticeship did not always suit Chandlee, as Jackson reprimanded him in April 1769: 'I do not approve of thy calling apprenticeship, slavery; perhaps thou wilt not consider how happy thou hast been 'till thou hast much more care upon thy head.'[25] Besides their shared experiences as emerging tradesmen, the letters demonstrate a detailed knowledge of astronomy and the ability to make calculations concerning the position of celestial bodies. They also cast light on networks of exchange facilitated by the periodical press in this period. Given his trade, Jackson had a particularly detailed grasp of the market for periodicals and good access to a wide range of these publications.[26] Together, these facets of the correspondence are revealing about eighteenth-century urban, intellectual culture.

Some of Jackson's letters include the workings out of specific calculations, presumably for Chandlee to model, and also comparisons of other astronomers' reckonings and published figures. Typically, the letters are written conversationally but, occasionally, Jackson moves into a more didactic style, running through a particular concept or calculation for Chandlee. Jackson often signed himself 'Philalithes Astronomus', meaning lover of astronomical truth or, simply, 'Philalithes'.[27]

The letters also include comments on reading and updates on Quaker meetings, and Jackson would add jokes, riddles or aphorisms at the top of the page, designed to amuse. For example, in May 1769, this reassuring message appeared:

> Tho' plung'd in ills, and exercis'd in care,
> Yet never let the noble mind despair,
> For blessings always wait on virtuous deeds,
> And tho' a late, a sure reward succeeds. (Unknown)[28]

Occasionally, small diagrams would appear, such as the example from June 1769 shown in Figure 4.1.

Whilst the balance of subjects varies letter to letter, astronomy certainly takes the lion's share of the page, with Jackson commenting in the summer of 1769, 'But Astronomical Matters are finished, they having very well fill'd up my epistle, which otherwise would have been but short, for want of something requiring answers in thine, or other entertaining matter in mine.'[29] Thomas Chandlee was not Jackson's only student, although Jackson credited Chandlee with encouraging his own astronomical investigations and claimed to 'take more pains with thee then any other of my pupils'.[30] Despite the teacher–student relationship, Jackson fostered Chandlee's independence and even disavowed the need for a tutor in this field of enquiry:

> Don't be discouraged that thou hast not a tutor to hand always, for I know by my small experience and many others have known it (I believe) that astronomers may learn most of the science with[ou]t a teacher, else what had some of the most famous astronomers done, who learned many times, no doubt, what no other living men knew.[31]

In 1768, a letter noted Jackson's approval of Chandlee's progress – 'Thus I have finished my instruction astronomical, by aquainting thee with Parralexes' – and declared Chandlee no longer a scholar 'but a tyro [novice] astronomer'.[32] Jackson also noted Chandlee's

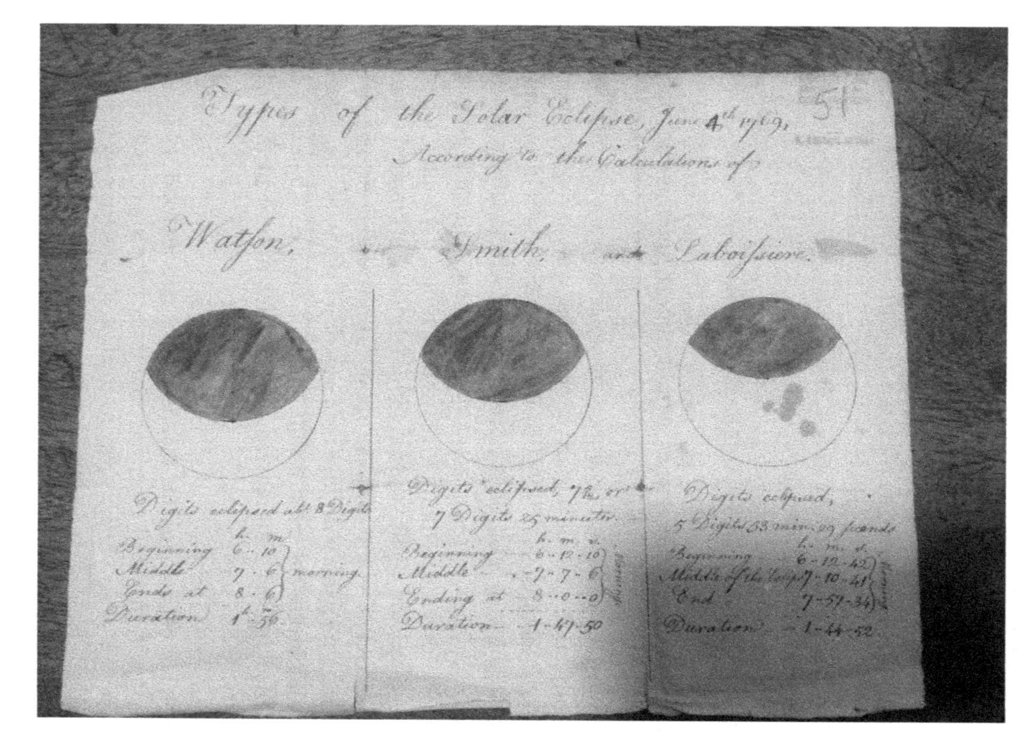

Figure 4.1 'Types of the Solar Eclipse', Robert Jackson, 4 June 1769.

superior eyesight and ability to distinguish objects at a greater distance than himself – suggesting, modestly, that Chandlee was in fact the better astronomer.[33] He even addressed letters to 'T. Chandlee, Bridestreet Astronomer-Royal', drawing his friend into a tongue-in-cheek aggrandisement of their shared endeavour.[34] Nevertheless, throughout this correspondence, Jackson posed questions to Chandlee – testing his capacities of observation and calculation – and expected them answered by return of mail.

Astronomy on a shoestring

It is tempting to imagine that only the deep pockets of wealthy landed gentlemen or aristocrats could give rise to the extensive use and regular adaptation of domestic space for intellectual work, but this was not the case. Similar practices, on a more modest scale, are visible in the correspondence of Jackson and Chandlee. Both young men lived in relatively limited domestic space, further compromised by the presence of other people. Their time was dominated by the demands of their respective apprenticeships. Despite these hindrances, they found room for their favourite occupation.

Robert Jackson lived in his father's premises – a house and shop – alongside other apprentices. In these busy surroundings, he was able to make himself a small study space that he referred to as the 'Hygrometer closet' on account of its containing such an instrument.[35] Jackson made active use of the space available; for example, he described two methods for making a meridian line, one of which used the shadows cast by a casement window on the floor of a room.[36] Another letter speaks to the chance sightings possible within even confined domestic space, as Jackson mentions seeing Saturn as he was going upstairs on 3 December 1769.[37] On occasion, to gain an improved view of the 'Western side', Jackson craned out of a 'back Garrett window', which he described as 'my best Uraniburg' in reference to the sixteenth-century Danish observatory of the same name.[38] Regular notes appear in these letters about the specifics of views possible from, often, the top windows in their respective city homes. One evening, Jackson enquired, 'Hast thou seen lucida lyra peeping late over the houses (not yet to be seen from the street but from a window) towards the N. East?'[39]

This example shows that eighteenth-century investigators used their homes flexibly, pushing their spatial and material affordances to accommodate a wide range of activity, even when other members of their household had different designs on the space.

Another major obstacle to scientific enquiry was the lack of access to reliable instruments of measurement. However, astronomy could be undertaken with a few basics or the possibility of borrowing a friend's apparatus. By and large, Jackson's access to suitable equipment and space for this pursuit out-stripped Chandlee's, but he was often generous with his resources. In November 1768, Jackson offered to lend Chandlee his quadrant, an instrument capable of measuring altitude.[40] However, Jackson stressed that to make use of the gadget, Chandlee would need to be shown how to operate it in person, emphasising the relevance of embodied knowledge.[41] A year later, Jackson suggested Chandlee might borrow his pocket quadrant to enable measurements on the hoof, and he asked that his friend use a watch to note the time of the sighting and report these and other relevant details back, allowing Jackson to discover which precise star he had seen.[42] Despite the centrality of the telescope to astronomy in this period, 'it by no means displaced sextants and quadrants' and 'telescopic sights', the latter 'arguably contributing more to astronomical observations' than the telescope at this time.[43]

Less easy-to-loan items were still made available to others, as Jackson hoped 'soon to have the hygrometer ready for your inspection & Isaac's or Thomas's if I could catch him at some conven[ien] t time'.[44] In December 1769, Jackson wrote to say that he had made Chandlee a pocket calendar designed for astronomical annotations and notes.[45] In another letter, Jackson described – for Chandlee's benefit – his own approach to annotating almanacs with observations.[46] These examples, including a homemade note-keeping technology, underline the prevalence of record-keeping practices in manuscript, print and hybrid formats discussed in Chapter 2.

As described, these young men lived in busy households. Their references to garret windows, attic spaces and closets signify the position of their own sleeping quarters in the least salubrious parts of the home. However, Jackson had access to another more specialised space for studying the skies – Crumlin House – which he described as a 'lodging' and 'a convenient empty house of 5 or 6 rooms'.[47] This site was located about two miles southwest of Dublin and was made available to a group of astronomers for making observations.

On 7 August 1769, Jackson reported there being '4 folks' who 'lie [here] ... every night as yet, of which 2, and I one of the 2; depart every morning, & return in the Evening'.[48] Jackson's use of Crumlin House certainly served his purposes. In a letter written partly at Crumlin and partly at Meath Street he recalled a series of sightings: 'At Crumlin one night since my last, I saw Charleswain, Arcturus, 2 shoulders of Auriga, Bootes, the Polestar, most of Swan ... Lyra, almost all the Dragon & Cassiopeids Chair.'[49] Jackson had also seen 'Caroli between the horizon & the last star of the Bear's tail' but failed to glimpse 'Spica', 'Antares, Mars, the Pleiades' and Venus.[50] As these observations suggest, the apprentices were attempting to gain sight of some of the brighter stars and planets in the sky, and to identify key constellations.

No doubt, Crumlin House's location outside of the city reduced some of the interference from residual urban light and thereby increased the likelihood of a clear sighting. Jackson referred to the difficulty of seeing clearly in the city in a letter written on 12 August 1769: 'I understand thou art by this time in town again; I have not much to write about now, What observations hast thou made in a place where the smoky Canopy of Dublin could not dim the stars to thy view?'[51] One letter, addressing Chandlee as 'Astro Professor', referred to a star being 'visible from thy observatory' meaning Chandlee's home in Bride Street, but obscured from his vantage point in Crumlin House: 'I can't expect well to see him, the southern part of my horizon is so encumber'd with a steeple, trees &c. I have a better chance to see him from Meath Street.'[52] A postscript on the same letter revealed Jackson back in situ in the Meath Street hygrometer closet, perhaps hoping for a glimpse.[53] However, for largely superior views of the night sky, Jackson made the journey on foot between Meath Street and Crumlin regularly, each leg taking him '42, 43 or 44 minutes'.[54] This comment speaks to Jackson's inclination for taking repeated measurements with his watch, but also to the importance of slivers of time that could be used as he pleased.

Although it was Jackson alone who had ongoing access to Crumlin House, he revealed the urge to share this with Chandlee when he wrote, 'I want thee at Crumlin to see the Garden ... I was thinking if we could appoint some time, suppose 1st day evening to meet ab[ou]t 6 and walk thither. Fine views of the milky way now at night at Crumlin.'[55] Jackson's use of an alternative space for astronomical observation was unusual; most working people who were curious about the natural world had to conduct their investigations within the spaces of home. Moreover, domestic spaces for scientific enquiry were often plagued by disruptions; Jackson reported on 22 October 1768 that 'all things (the hygrometer and a few others excepted) have been turned out of the Hygrometer closet & the room adjacent. My ill-looking desk was whirled into the dining lumber room, from whence I now write.'[56] Nonetheless, the home re-imagined as an observatory and pushed to the limits of its spatial affordance was where Jackson and Chandlee set to in their mutual investigation of the night sky.

A community of astronomers

In May 1769, Jackson wrote to Chandlee to ask, 'Hast thou observed a star not far west of Jupiter called the South Balance, and over him (Jupiter) another in the middle of the beam, both these bright stars of the 2nd magnitude – about 2 hours after Jupiter rises Antares[?]'[57] In this way, the letter-writers exchanged details of their independent observations, sharing in the excitement of sightings together. However, these apprentices were in touch with a network of other astronomers across the city, and aware of a broader community of star-gazers – including famous individuals – through their reading of periodical literature. For example, on 17 April 1768, Jackson commented, 'The Empress of Russia & her astronomers are ab[ou]t preparing to observe the transit of venus, tho' so far off, as 6th mo. [June] 1769.'[58] On 22 October of that same year, Jackson discussed astronomers including Charles Leadbetter (1681–1744), in particular his Table of Eclipses for the years 1724–40.[59] Another letter gossiped about a disagreement between fellow Dublin star-gazers:

> This is but dull sort of Weather for observing a transit of Venus. I heard that some months ago there was a dispute between Calcearius [shoemaker] of George's Lane & a Ludimagister [teacher] in Meath Street. Calcearius asserted that a certain firy-looking [*sic*] star that had then lately been seen in Conjunction with the Moon, was the Planet Mars; but Pedagogic [teacher] affirmed that it must be Saturn, whereupon an Ephemeris[60] being got, it decided it in favour of Calcearius [shoemaker].[61]

These comments illuminate several characteristics of Jackson's engagement with astronomy. Much as he used a pseudonym – 'Philalithes Astronomus' – inspired by Ancient Greek, the use of Latin words to describe local contacts signalled a familiarity with Classical languages, if only a passing one. Certainly, the humorous performance of his own role of teacher in his friendship with Chandlee seems a more likely motivation than the maintenance of the anonymity of a neighbourhood shoe-maker and teacher. This letter also underlines the inclusive nature of astronomical enquiry, encompassing tradespeople and professionals and, in this case, people known to Jackson personally or by reputation. The shoemaker's address was most likely on the northern end of modern-day South Great George Street, next to Dublin Castle, whereas the teacher lived in the same street as Jackson – part of the Liberties, an area with a thriving textile industry at its peak in the late eighteenth-century city. The fact the disagreement was settled by consulting an 'ephemeris' – a table providing details of the positions of planets and stars over a period of time – emphasises the centrality of print culture in enabling 'amateur' astronomy at this time.

These letters also document regular meetings of like-minded acquaintances on St Stephen's Green to observe the stars. In one letter, Jackson described doing so early in the morning before he started the day's work and, in another, he declared the season too cold and himself too busy for such a rendezvous with Chandlee.[62] On 7 June 1769, Jackson wrote warmly to his 'Loving Correspondent' to report an exchange between himself and two other astronomers: 'Hutchinson communicated to me his Observations on the Eclipse, which were much preferable to either Harding's or mine.'[63] Hutchinson earns several mentions in these letters, sometimes referred to as the 'astronomer of High Street'[64] and also as Jackson's 'astro brother'.[65] However, a decent timepiece had aided Hutchinson's measurements:

> he found by his well regulated clock (adding 2 min to make it agree with the apparent or solar time) that the Eclipse began at 6:13 and ended at 7:59, digits not exactly measured, but he found that the Moon's edge went beyond the sun's Centre, which shews it to have been above 6 Digits.[66]

By collecting this information from a range of independent viewers, Jackson was able to critique the calculations printed in periodicals: 'I think many of the Almanack writers were much mistaken about it'; one 'made the Digits somewhat too small' and another 'erred in making the duration near 10 min too much'.[67]

In June 1769, Jackson commented on another discrepancy between the calculations of Dublin-based astronomers, which had taken place several decades ago: 'but this is but very little in comparison of what happened in 1737, when T. Hutchinson observ'd the great solar eclipse and found it to differ 28 min from the Calculation of an Astronomer that lived not many miles from Chequer lane'.[68] Later that year, Jackson noted further criticisms of the astronomical content in Watson's Dublin almanac.[69] In these comments, it is also clear that Jackson was not only concerned with current or very recent astronomical activity but tracked back decades to understand the context for his own activities and those of other astronomers, famous or otherwise.[70] This concern with the long-term was characteristic of astronomy, which relied on centuries-old record-keeping to determine celestial cycles.[71]

On 7 June 1769, Jackson wrote to Chandlee concerning a partial eclipse of the sun that had occurred first thing on the morning of Sunday 4 June and also a transit of Venus that had taken place over 3–4 June. The latter represented an important step forward in human understanding of the universe because both the 1761 and 1769 transits of Venus (witnessed as a small black disc travelling across the surface of the sun) offered astronomers across the globe the opportunity to make measurements that, taken together, could confirm an accurate distance between the earth and the sun. Jackson communicated a range of sightings from among his network, including an R. Blood and John Wilcox seeing the transit 'with prospect glasses', one from 'Clibborn's terrace', which likely referred to a property in Banbridge, Co. Down, owned by the Quaker Clibborn family, and the other from 'the bank house'.[72] However, 'the most curious observation on the transit was made by a lad of my acquaintance in Meath Street; who with (I suppose) a very bad piece of glass smoked, saith he saw something very near the sun, with a tail to it'.[73]

Jackson was also keen to hear more about the company Chandlee was in when he 'led [them] forth into the Park to observe the eclipse'.[74] The urgency of capturing as much information as possible about the eclipse rested on its relative rarity: 'We had never a visible eclipse of the sun since 1766 to this late one, and I imagine we will not see another at least these two years.'[75] Jackson seemed disappointed with the limited use Chandlee had made of this unusual opportunity, remarking, 'I think if thou hadst been diligent thou might have found the duration to less than a minute.' However, he had to recognise the constraints of Chandlee's setup, acknowledging that 'the help of my mock telescope' might have made such a calculation more possible.[76] Whilst Jackson could not afford the expense of an instrument-maker's telescope, this reference to a 'mock telescope' suggests he had constructed something himself to approximate the effect. This interpretation is underlined by a comment in another letter:

> Our glasses must be very different for I looked at Mars with a Tube which ought to have 3 glasses but only one remains; and he appeared like a large pin's head, of a fine colour, Jupiter and the fixt stars in like manner had their apparent magnitudes lessened by it.[77]

Not only were some of Jackson and Chandlee's tools homemade, but they were also incomplete or failing. Nevertheless, the correspondence shows a determination to continue regardless and to incorporate the fallibility of the instruments – their own and other people's – into the analysis of what had been seen and, therefore, what could be known. However, this zeal was notably absent in other members of Jackson's household, as he mentioned 'Many lay in bed and saw not the Eclipse'; besides himself, only one other member of the Meath Street house witnessed the phenomenon: 'thus laziness hindred many of seeing what I think deserved notice'.[78] Before concluding his letter, Jackson commented, 'The Crumlin House was admirably well situated for observing the transit, but to my regret not so for the eclipse, for seeing which the Meath Street house was tolerably well situated.'[79]

These two intense years of apprenticeship and astronomy came to a close in 1769, when – ahead of Chandlee – Jackson's indenture expired and he became a journeyman. In a letter dated 23 April 1769, Jackson reported tidying up the house before his departure: 'last week & this week, house & shop to be brushed up and put in order, so I have something to do but that is not wonderfull as it is the Case every week'.[80] However, he could not end the letter without a brief comment on the stars and an offer of help: 'Mars has now got as far as Castor's foot – will not the representation of nocturnal appearance of 5 mo. 1st [1 May], be soon of use to thee?'[81] Then Jackson signed off: 'Farewell, I remain thy Wellwisher and old Acquaintance, Astronomus.'[82]

It seems certain Jackson and Chandlee remained in touch thereafter, but this period of regular corresponding casts light on the incredible curiosity of both men and the determination with which they developed their interest in astronomy and the skills they could bring to bear in both calculation and observation. It is worth noting that through this period of frequent letter exchange, the men also had the opportunity to meet in person – to confer on the particulars of a given exercise or to make a sighting together, alongside others of their local acquaintance, on St Stephen's Green. Despite substandard instruments and a heavy daily workload that took them away from their windows and periodicals, the depth of engagement was significant – their knowledge was considerable.

Conclusion

Here, Jackson and Chandlee can be seen to illuminate important features of the culture of curiosity in this period, despite occupying marginal positions in relation to the institutions and high-profile personalities of Enlightenment science. It is also worth noting that regardless of the vast disparities in financial resources, domestic space and family connections that existed between the working men in this chapter and Lady Clive, discussed in Chapter 3, their level of engagement was not dissimilar. Both apprentice and aristocrat struggled to see some celestial bodies through the instruments at their disposal, they both relied upon published tables and print culture to compare and contrast findings and, in each case, the corroboratory information made possible by networks of astronomers, from many walks of life, fuelled their enquiries.

The testimony of Jackson's letters brings themes to the fore that are important for a full understanding of the experience and practice of 'science' in eighteenth-century Ireland and Britain. Ursula Klein has identified 'bodily skills to connoisseurship of materials, tacit and verbal to articulated know-how, to methods of measuring, data gathering, and classification', analysis and representation as important skills and knowledge for eighteenth-century scientists – all of these qualities and activities can be seen at work in the letters discussed here.[83] Likewise, as Daston has also characterised for *virtuosi*, scientific observation for Jackson and Chandlee regulated their lives, from the routes they walked across the city, to the repeated cycles of observation, calculation, reference and comparison. Astronomy also shaped these men's social lives and networks of association.[84] Their letters offer a glimpse into tradespeople's households and the ways in which these spaces and their regimes of labour drove and shaped enquiry.

Notes

1 Dear, 'Meanings of experience', p. 108.
2 *Ibid.*, p. 130; see also Peter Pesic, 'Wrestling with Proteus: Francis Bacon and the "torture" of nature', *Isis*, 90 (1999), pp. 81–94.
3 Daston and Lunbeck, *Scientific observation*, p. 1; Lorraine Daston also describes it as 'a key learned practice' and a 'fundamental form of knowledge' that had come of age by the mid-eighteenth century, see 'Empire of observation', p. 81.
4 Daston and Lunbeck, *Scientific observation*, p. 3.
5 *Ibid.*, p. 86.
6 *Ibid.*, p. 99.
7 Anne Secord, 'Introduction' in G. White (A. Secord (ed.)), *The natural history of Selbourne* (Oxford: Oxford University Press, 2013), p. xvii.
8 *Ibid.*, p. xxiii.
9 White, *Natural history of Selbourne*, p. 17.
10 *Ibid.*, p. 16, n. 1.
11 Secord, 'Introduction', p. xviii.
12 See Daston, 'Empire of observation', pp. 99–100 on repetition in observation.
13 Secord, 'Introduction', p. xx.
14 Brück, *Women*, p. xv.
15 Dear, 'Meanings of experience', p. 121.
16 Capp, *Astrology*, p. 191.
17 *Ibid.*, p. 15.
18 *Ibid.*, p. 23; in the 1660s, approximately one in every three English households purchased an almanac annually.
19 Charles Benson, 'The Irish trade' in Suarez and Turner, *Cambridge history of the book*, pp. 366–82.

20 Shelley Costa, 'The "Ladies' diary": Gender, mathematics and civil society in early-eighteenth-century England', *Osiris*, 17 (2002), pp. 49–73; despite being aimed at women readers, the readership was comprised of both sexes.

21 Angela N. H. Creager, Mathias Grote and Elaine Leong, 'Learning by the book: Manuals and handbooks in the history of science', *British Journal for the History of Science: Themes*, 5 (2020), p. 9 (pp. 1–13); see also Ursula Klein, *Experiments, models, paper tools: Cultures of organic chemistry in the nineteenth century* (Stanford, CA: Stanford University Press, 2003).

22 On Isaac and Robert Jackson's careers in printing, see Pollard, *Dictionary*, pp. 311–15; father and son had close connections with other major Quaker figures of the period, including John Rutty (1698–1775), Isaac having printed several of his works and Robert acting alongside other notable Quakers as executor to Rutty's will. See Richard S. Harrison, *Dr John Rutty (1698–1775) of Dublin: A Quaker polymath in the Enlightenment* (Dublin: Original Writing, 2011), pp. 163, 173, 178, 224.

23 A trained worker who is employed by someone else.

24 Prior to Robert's tenure as apprentice, his father had taken on at least three other apprentices, one William Stroud who ran away in 1744 and another Anthony Harman who was bound from Bluecoat School in November 1760; see Pollard, *Dictionary*, p. 312.

25 Friends Historical Library Dublin (hereafter FHLD), Selina Fennell Collection (hereafter Fennell), MSS Box 27, folder 1, letter 19: Robert Jackson to Thomas Chandlee, 9 Apr. 1769.

26 Distinguishing between newspapers and periodicals is difficult in this period. Here, the term periodical will be used to denote the large category of publication that was printed on a regular basis and, often, in a cheap format – at least as compared with books. See Tierney, 'Periodicals and the trade', p. 479.

27 The Ancient Greek name, 'Philalethes' (modern spelling), was often used as a pseudonym on account of its meaning 'lover of truth'.

28 FHLD, Fennell, MSS Box 27, folder 1, letter 22: Jackson to Chandlee, 7 May 1769.

29 FHLD, Fennell, MSS Box 27, folder 1, letter 25: Jackson to Chandlee, 23 Jun. 1769.

30 FHLD, Fennell, MSS Box 27, folder 3, letter 105: Jackson to Chandlee, n.d.; the use of the old-fashioned terms 'thee' and 'thou' was typical for Quakers in this period.

31 FHLD, Fennell, MSS Box 27, folder 3, letter 123: Jackson to Chandlee, n.d.

32 FHLD, Fennell, MSS Box 27, folder 3, letter 91: Jackson to Chandlee, n.d. (but likely 1768).

33 FHLD, Fennell, MSS Box 27, folder 2, letter 79: Jackson to Chandlee, n.d.
34 FHLD, Fennell, MSS Box 27, folder 3, letter 87: Jackson to Chandlee, n.d.
35 See, for example, FHLD, Fennell, MSS Box 27, folder 1, letters 56 and 75; folder 3, letters 97, 99 and 110. A hygrometer is an instrument used for measuring the amount of humidity and water vapour in the atmosphere, in soil, or in confined spaces. See Mateus, 'Searching', p. 163.
36 FHLD, Fennell, MSS Box 27, folder 2, letter 52. However, in folder 3, letter 99: n.d., Jackson noted, 'But it's likely thou are not possessed of a room convenient to do it in. So I may spare my labour.'
37 FHLD, Fennell, MSS Box 27, folder 1, letter 31: Jackson to Chandlee, 3 Dec 1769.
38 FHLD, Fennell, MSS Box 27, folder 3, letter 85: Jackson to Chandlee, 3 May *c.* 1768.
39 *Ibid.*; 'lucida lyra' most likely refers to Vega – the brightest star in the northern constellation of Lyra.
40 The quadrant takes angular measurements of altitude and is usually comprised of a graduated quarter of a circle with a mechanism for sighting. Quadrants were used for both astronomy and navigation in this period.
41 FHLD, Fennell, MSS Box 27, folder 3, letter 82: Jackson to Chandlee, 25 Nov. 1768.
42 FHLD, Fennell, MSS Box 27, folder 1, letter 31: Jackson to Chandlee, 3 Dec. 1769.
43 Daston, 'Empire of observation', p. 94.
44 FHLD, Fennell, MSS Box 27, folder 3, letter 87: Jackson to Chandlee, n.d.. Similarly, arrangements are made for Chandlee to see the hygrometer in folder 3, letter 80: Jackson to Chandlee, 12 Nov. 1768.
45 FHLD, Fennell, MSS Box 27, folder 1, letter 33: Jackson to Chandlee, 15 Dec. 1769.
46 FHLD, Fennell, MSS Box 27, folder 3, letter 125: Jackson to Chandlee, n.d.
47 FHLD, Fennell, MSS Box 27, folder 2, letter 76: Jackson to Chandlee, 31 Jul. *c.* 1768.
48 FHLD, Fennell, MSS Box 27, folder 2, letter 56: Jackson to Chandlee, 7 Aug. *c.* 1768. Olive C. Goodbody notes the use of Crumlin House by a group of astronomers and suggests that this might be a general reference to a property in Crumlin; see *Guide to Irish Quaker records, 1654–1860* (Dublin: Stationery Office for the Irish Manuscripts Commission, 1967), p. 64.
49 FHLD, Fennell, MSS Box 27, folder 3, letter 81: Jackson to Chandlee, 21 Aug *c.* 1768. 'Charleswain' is a bright, circumpolar asterism;

'Arcturus' is one of the brightest stars in the northern hemisphere and is found in the 'Boötes' constellation; 'Auriga', 'Swan' (Cygnus), 'Lyra' and 'Dragon' (Draco) all refer to northern constellations; and 'Cassiopeids Chair' refers to the five brightest stars in the constellation Cassiopeia.

50 FHLD, Fennell, MSS Box 27, folder 3, letter 81: Jackson to Chandlee, 21 Aug. *c.* 1768. 'Caroli' refers to Cor Caroli – a binary or double star; the 'Bear's tail' suggests the constellation Ursa Major; 'Spica' and 'Antares' are very bright stars, 'Pleiades' is a group of over 800 stars in the Taurus constellation; Mars and Venus are, of course, planets.

51 FHLD, Fennell, MSS Box 27, folder 1, letter 26: Jackson to Chandlee, 12 Aug. 1769.

52 FHLD, Fennell, MSS Box 27, folder 3, letter 110: Jackson to Chandlee, 15 Sep. *c.* 1768.

53 *Ibid.*

54 FHLD, Fennell, MSS Box 27, folder 2, letter 56: Jackson to Chandlee, 7 Aug. *c.* 1768/9.

55 FHLD, Fennell, MSS Box 27, folder 1, letter 26: Jackson to Chandlee, 12 Aug. 1769.

56 FHLD, Fennell, MSS Box 27, folder 3, letter 97: Jackson to Chandlee, also folder 2, letters 56 and 75.

57 FHLD, Fennell, MSS Box 27, folder 1, letter 22: Jackson to Chandlee, 7 May 1769.

58 FHLD, Fennell, MSS Box 27, folder 2, letter 72: Jackson to Chandlee, 17 Apr. 1768. These kinds of reports were common in the periodical press in this period; for example, in the *Gentleman's Magazine*, vol. 10 (Feb. 1740), Mr I. N. De L'Isle, first astronomer to the Empress of Russia, receives a mention in relation to the 1739 solar eclipse alongside sightings submitted from 'I. B.' of Stoke Newington and 'J. T.' of Newcastle-on-Tyne, p. 80.

59 FHLD, Fennell, MSS Box 27, folder 3, letter 97: Jackson to Chandlee, 22 Oct. 1768; see also Charles E. Sayle (ed.), *A catalogue of the Bradshaw Collection of Irish Books in the University Library Cambridge, vol. 1, books printed in Dublin by known printers, 1602–1882* (Cambridge: Cambridge University Press, 2014), p. 199.

60 An astronomical term, meaning a table providing the calculated positions of a celestial object at regular intervals throughout a period.

61 FHLD, Fennell, MSS Box 27, folder 1, letter 23: Jackson to Chandlee, n.d.

62 FHLD, Fennell, MSS Box 27, folder 3, letter 87, n.d.; letter 82, 25 Nov. *c.* 1768.

63 FHLD, Fennell, MSS Box 27, folder 1, letter 25: Jackson to Chandlee, 7 Jun. 1769.

64 FHLD, Fennell, MSS Box 27, folder 3, letter 98: Jackson to Chandlee, 16 Oct. *c.* 1768, see also folder 1, letter 24.
65 FHLD, Fennell, MSS Box 27, folder 2, letter 69: Jackson to Chandlee, n.d.
66 FHLD, Fennell, MSS Box 27, folder 1, letter 24: Jackson to Chandlee, 7 Jun. 1769.
67 *Ibid.* Laboissière and Watson were the named almanac writers, the latter regularly mentioned in this correspondence; see also folder 1, letter 30, folder 2, letters 51 and 77.
68 FHLD, Fennell, MSS Box 27, folder 1, letter 24: Jackson to Chandlee, 7 Jun. 1769; Chequer Lane refers to a street located between Dublin Castle and Trinity College Dublin in the city centre referred to as Exchequer Street today.
69 FHLD, Fennell, MSS Box 27, folder 1, letter 30: Jackson to Chandlee, 18 Nov. 1769.
70 This behaviour was not unusual for periodical readers, as Gillian Williamson has identified for the *Gentleman's Magazine*; see *British masculinity*, p. 36.
71 Daston, 'Empire of observation', p. 93.
72 FHLD, Fennell, MSS Box 27, folder 1, letter 24: Jackson to Chandlee, 7 Jun. 1769.
73 *Ibid.*
74 *Ibid.*
75 *Ibid.*
76 *Ibid.*
77 FHLD, Fennell, MSS Box 27, folder 3, letter 99: Jackson to Chandlee, n.d.
78 FHLD, Fennell, MSS Box 27, folder 1, letter 24: Jackson to Chandlee, 7 Jun. 1769.
79 *Ibid.*
80 FHLD, Fennell, MSS Box 27, folder 1, letter 21: Jackson to Chandlee, 23 Apr. 1769.
81 *Ibid.*; the foot of Castor refers to part of the Gemini constellation, associated with the twins of Greek mythology, Castor and Pollux.
82 FHLD, Fennell, MSS Box 27, folder 1, letter 21: Jackson to Chandlee, 23 Apr. 1769.
83 Ursula Klein, 'The laboratory challenge: Some revisions of the standard view of early modern experimentation', *Isis*, 99:4 (2008), pp. 781–2 (pp. 769–82).
84 Daston, 'Empire of observation', p. 106.

5

Experimenting

The Latin terms for 'experience' and 'experiment' had been used interchangeably in medieval and early modern writings and by the beginning of the eighteenth century, 'the construal of experience as "experiment" ... had acquired a wide and influential currency'.[1] As Étienne Chauvin argued at the turn of the eighteenth century, 'reason without experience is like a ship tossing about without a helmsman'.[2] Taken together, observation and experiment were fundamental to the scientific developments of this era, and both required direct and personal experience of phenomena and the ability to record that process. As mentioned in the previous chapter, observation and experiment often went hand in hand, and the examples discussed here reveal elements of both. Alongside a recognition of experience as a route to understanding came the acknowledgement of artisanal knowledge as important to scientific enquiry and, with it, a greater value placed on 'useful knowledge' as part of the larger search for truth.[3] Much of what follows draws on the records of societies dedicated to developing such useful knowledge. However, as these case studies show, the home was also a primary space for experiment.

The main focus of this chapter is on the experimental work of breeding silkworms, and the central examples include a post-mistress in Kent, an apothecary in Pennsylvania and a range of other working and leisured women – all of whom conducted their experiments from home. Mary Terrall has explored the use made by naturalist René Antoine Ferchault de Réaumur (1683–1757) of domestic space for experiment and observation, noting his dependence on the capacities and personnel of his two large residences.[4] Here, elite domestic space is considered alongside homes of a much more modest scale – revealing that the large homes of the propertied classes were not a prerequisite for active domestic enquiry.

Mrs Wyndham's scientific life

To begin, however, here is an example that raises more questions than it answers, especially in terms of gender, class and intellectual agency. In 1796, Elizabeth Wyndham won a silver medal from the Society for the Encouragement of Arts, Manufactures and Commerce in London under the category of 'Mechanicks' (see Figure 5.1). Her innovation was a cross-bar lever, which she had designed to help resolve her workmen's difficulties in moving large and heavy rocks – or, as the Society put it, 'her ingenious contrivance of a method

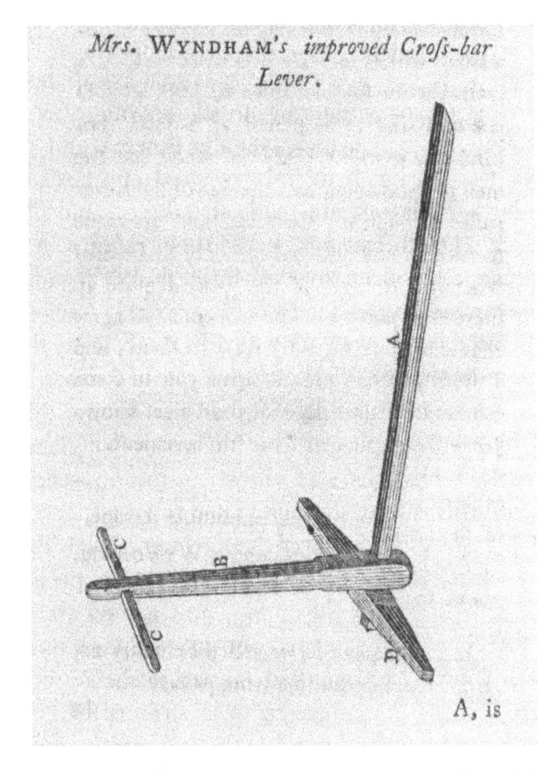

Figure 5.1 Mrs Wyndham's cross-lever. Courtesy of Royal Society of Arts, London.

of using to the best advantage, the power applied to the Cross-Bar Lever, for raising large weights'.[5] On 28 October 1795, Wyndham sent a drawing of the lever and a model (which was then stored in the Society's Repository for the Inspection of the Public) alongside her explanatory correspondence. She wrote, 'I have sent you a Model of a mechanical invention of my own, which you will laugh at, as every body here did at first; but I assure you, it has proved of great use, and the workmen all approve of it very much.'[6] However, when Wyndham had first observed the men making use of her lever, she noticed that they did so

> in a very ineffectual manner, by standing three or four at a time on the bar of the Lever by which means some of them were placed so near the fulcrum, that their power was in great degree lost; besides they were obliged to steady themselves upon sticks, for fear of falling, which took off from their weight upon the Lever.[7]

She explained how her invention was intended to be operated, cross-referencing the drawing and model showing how her design 'inclines backwards, which increases the power', and included 'a cross-bar for the workmen to hold by' and another for them to stand on – both 'additions are made to take on and off, and are only to be used when the strength of the rocks require an increase in power'.[8] Fortunately, once corrected in its usage, Wyndham noted that 'The workmen all agree that it is of very great service to them.'[9]

Elizabeth 'Wyndham' was actually Elizabeth Ilive, the mistress of George Wyndham, third Earl of Egremont of Petworth House in West Sussex, whose household accounts were discussed in Chapter 1. It is worth briefly exploring the life history of Elizabeth Ilive (*c.* 1770–1822). Whilst she was known in the household as 'Mrs Wyndham' and this is how she signed her correspondence with the Society in 1795, she did not marry the Earl until 1801. By this time, Ilive was over thirty, the Earl nearly fifty and the couple already had seven children. An article by Alison McCann, former archivist of the West Sussex Archive, first sketched out this woman's unusual story back in 1983.[10] What McCann had spotted in the Petworth House Archive were records that suggested a laboratory had been established in the house towards the end of the eighteenth century. Whilst it was not uncommon for rooms in large households to be adapted for scientific purposes, these operations tended to be undertaken by the master or mistress of the household and it is interesting that in this case a woman of modest origins and in an insecure position as a mistress, not a wife, nonetheless exercised her will to make her home a haven for scientific enquiry and discussion.[11] Indeed, at a neighbouring estate, Goodwood House in Sussex, Charles Lennox, third Duke of Richmond (1735–1806) took the initiative to create his own laboratory in 1790. However, in the case of Petworth House, it is Ilive's work that is visible in the records.

Relatively little is known about Elizabeth Ilive's background, and she has been reported as being either the daughter of a librarian at Westminster School or the daughter of a Devon farmer. It is thought that she met the Earl when she was a teenager in 1786 and her first child by Egremont was born the following year in 1787. Owing to her long-term position as an unmarried mistress, Ilive's life at Petworth House had to accommodate inherent difficulties of status and authority. When the company was composed of family and visiting artists, Ilive was permitted to dine downstairs, but when visitors of higher social standing were received she would not form part of the gathering.[12] The flow of artists and writers that came through Petworth House provides further evidence of Elizabeth Ilive's active encouragement of intellectual discussion within the household; she was – for example – the patron of several artists including William Blake, who dedicated to her his painting *The last judgement*.[13]

Ilive clearly maintained her own artistic and scientific interests at Petworth House, but the Earl was also motivated by intellectual concerns. As discussed in Chapter 1, Petworth household accounts reveal considerable resources put to the service of art and science. Egremont is known to have supported J. M. W. Turner with his patronage and, like many great landowners of his era, he followed scientific developments in the field of agriculture and husbandry particularly closely.[14] Examples of the Earl's purchases in the 1770s and 1780s point to wide-ranging interests: he acquired Joseph Priestley's *The history and present state of electricity* (first published in 1775) and a few years later he bought some electrical machines for the estate. He also secured a two-foot telescope from an optical and mathematical instrument-maker; he was a patron of the Vaccine Board and established a surgery and dispensary at Petworth between 1789 and 1790. The Earl's adaptations to the

Petworth estate also included the creation of a museum in the late 1790s, which he put in the hands of a naturalist, Reverend Robert Ferryman (1753–1837).[15]

Despite the prominence of the Earl's intellectual concerns, the household accounts make it possible to determine Ilive's purchases of this nature. In many cases, she used prestigious London suppliers for instruments, glassware and chemicals. An iron furnace was also ordered especially and sent from the Norwich Iron Foundry of J. Peckover and fitted in the week of 24 March 1798.[16] Subsequently, a chimney was raised and fires put in to service the laboratory. Many of the instruments, vessels and materials ordered for the laboratory were likely to have been used in the demonstration of the powers of vacuum – a phenomenon that could easily have been shown to others visiting the house. Other retorts and chemicals recorded in the accounts might have been used for chemical experiments, although it is difficult to be specific about the exact experiments that Ilive conducted in her laboratory.

It has not been possible to locate where exactly the laboratory was built in Petworth House, although a collection of eighteenth-century scientific equipment remains in a cupboard in the household.[17] To date, no evidence has even been uncovered of the furnace, which would have had the largest footprint on the space. In fact, one of the only qualitative records of Elizabeth Ilive's contribution to matters scientific remains the silver medal from the Society for the Encouragement of Arts, Manufactures and Commerce discussed above.

Ilive's work in the laboratory was to be short-lived. After her marriage in 1801, the couple's relationship deteriorated sharply, and they lost their only legitimate child the following year. In May 1803, a deed of separation was drawn up and Elizabeth, now Countess of Egremont, moved out of Petworth House – she would not return. Little is known of her life thereafter. Whilst Ilive herself remains a relative enigma, the fragments of evidence point to scientific activities conducted on an extraordinary scale, especially considering her marginalised position within an aristocratic household. Although very little of her own testimony survives to place alongside the evidence of account books and glass vessels, other curious women were voluble on the subject of their own intellectual activity, as what follows will show.

Silkworm breeding

An activity that sat at the juncture of naturalism and commercial interest, silkworm breeding was taken up by many women across Britain and Ireland in this period. Championed by learned societies, institutional records attest not only to widespread participation in this pursuit but also to the intricacies of observation and experiment as they took place at home.

In the eighteenth century, silk was produced almost entirely abroad but remained in high demand by consumers at home who acquired this product at considerable expense. As such, silk production offered an opportunity for innovation that could lead to the development of a new and profitable domestic (in both senses) industry. Since the early 1600s, there had been an appetite in England for this venture but it had been slow to develop.[18] In fact, seventeenth-century settlers in North America had been cheered by the recognition that the preferred food of silkworms, mulberry bushes, grew well in this region. As a consequence, they had foreseen a flourishing silk industry, but it wasn't until the eighteenth century that efforts became more coordinated to plant the correct variety of mulberry plants for this purpose.[19] The cultivation of mulberry bushes extended across the British Empire, with a J. Marten of Palamcotah, Madras, in India noting in 1791 'they are not uncommon in this country'.[20]

In Britain and Ireland, meanwhile, silk weaving alongside other textile production had become well established in the Liberties area of Dublin and Spitalfields in East London, although the raw material was still mainly imported.[21] Two new eighteenth-century organisations, the Society for the Encouragement of Arts, Manufactures and Commerce in London and the Dublin Society took up this cause.[22] The Dublin Society was granted its royal charter in 1820, becoming the Royal Dublin Society, and the London Society became the Royal Society of Arts in 1908 – these are the names the societies go by to this day. Both societies were focused on promoting 'useful knowledge' in a dizzying array of domains. To this end, prizes were offered for submissions by members of the general public that could shed light on all manner of questions, from the domestic production of chip hats to the planting of turnips.[23] There were two types of prize: 'premiums' for entries that responded to a call issued by the

Society and 'bounties', which were awarded to unsolicited submissions. The categories were:

Agriculture (which might concern the growing of vegetables; tree planting; sowing techniques or new farming technologies)
Manufactures (for example, improvements in techniques for dying leather; ways of manufacturing milled hats in imitation of the French; loom-woven fishing nets)
Chemistry (a wide-ranging category, including perfume production and food innovation among many other projects)
Mechanics (focused on innovation in technology – although these innovations were often put to agricultural uses)
Polite arts (this category included drawing; decorative arts; the development of paints and pigments; and experiments with natural dyes)
Colonies and trade (this covered initiatives such as vines being transported from the Old World to the New and the collecting of botanical samples).

The London-based Society was not the first of its kind and its establishment was prefigured by the Dublin Society, in 1731, and also the American Philosophical Society, which was initiated in 1743 by the polymath, Benjamin Franklin (1706–90), who later became one of America's founding fathers. All three societies built on the work of smaller philosophical associations that had proliferated across the British world since the last quarter of the seventeenth century, and which often attended to the ways science could be applied to practical problems.[24]

Like many of the other subjects of interest to these societies, the production of silk was really about import substitution. Where Britain and Ireland remained dependent on expensive foreign products, efforts were made to produce substitutes at home. However, in many cases efforts were thwarted by the variables of climate, ingredient or technique. The London Society's records note that whilst 'propagating Silk Worms, and obtaining Silk in England, was an early Subject of the Society's consideration', sourcing sufficient mulberry leaves to feed silkworms remained a major challenge.[25] In 1768, the London Society offered premiums for activities relating to silk production, including for 'the greatest quantity of merchantable Silk', 'an account of the best method of breeding and treating Silk Worms, in order to the obtaining Silk, verified by experiments' and 'For raising the greatest number of white or black Mulberry Trees'.[26] Between the years 1768 and 1790, there were nine formal calls for submissions relating to silk, which would have been advertised in the print press as well as through the Society's own communications and publications.[27]

By contrast, the Dublin Society published papers on the subject but did not issue any premiums, instead relying on ad hoc submissions that they might choose to reward. Nonetheless, works of advocacy for domestic silkworm experimentation were published in 1750 and 1799, revealing the longevity of the Dublin Society's interest in this topic.[28] On both islands, the domestic cultivation of silkworms capable of producing silk for the home market was enthusiastically taken up by interested individuals.

When householders ventured on this task, they engaged with zoology, botany, technology and matters of business, considering the cultivation of both worms and the mulberry trees upon which they depended alongside issues of equipment and labour that would affect the scale and profitability of this enterprise. As Reverend Samuel Pullein commented to the Dublin Society in 1750, it was thought that 'many thousand Spinsters of a more curious Nature, without the Expence of Wages' could become the workforce for this new silk manufacture and by doing so 'be of publick Good to their Country'.[29]

The mission of putting 'useful arts' to the service of the national good was at the heart of both of these eighteenth-century institutions. The Dublin Society was established to conduct philanthropic work 'to promote and develop agriculture, arts, industry and science in Ireland' and was founded by members of the Dublin Philosophical Society, principally Thomas Prior and Samuel Madden. The London Society was founded by the drawing teacher and inventor William Shipley to encourage creative thinking that could be put to public use. Both societies distinguished themselves from other intellectual institutions, such as the Royal Society, through their focus on the practical application of new knowledge. The Dublin Society emphasised the mission to provide 'useful' knowledge as opposed to 'laboured speculations' or the enrichment of 'the learned world'.[30] They wished instead 'to direct the industry of common artists; and to bring practical and useful knowledge from the retirement of closets and libraries into public view'.[31]

Whilst the founding members of both societies were drawn from the landed, professional and merchant classes, the memberships they amassed were more diverse.[32] By 1764, only 10 per cent of the London Society's 2,136-strong membership were individuals with titles and under 10 per cent were designated as medical, clerical, naval or military in background. Many more members were artisans, manufacturers, farmers or traders, including substantial representation from watch-makers and printers.[33] The Dublin Society had a much smaller membership numbering 267 in 1734. Not all of those named on the membership list played an active role and so a limited membership of 100 committed individuals was established thereafter.[34] Beyond the membership was an even wider section of the population who submitted proposals and designs to the societies, including those whose literacy was extremely basic.[35] Women existed within their number and a recent analysis of the London Society's holdings (1755–1852), across all categories (excluding the Polite Arts which incorporated premiums intended for women), found just over 3 per cent of submissions were penned by women.[36] Both societies were certainly committed to making their calls for submissions accessible to as many as possible and some of their ventures were explicitly aimed at women.[37] In fact, the societies' archives trace the remnants of a large, international network of correspondents. These diverse submissions provided 'rich particularity and local detail' – a feature of the London Society's work that was both valued and criticised in this period.[38] The work of these geographically and socially disparate individuals survives in part as original letters sent to the London Society (unfortunately none survive for the Dublin Society) and as recorded within the organisations' proceedings, minutes and transactions.

Female home experimenters in Britain and Ireland were drawn to the challenge of silkworm breeding, perhaps encouraged by the pamphlets of Pullein and others arguing that this was a pursuit that could be conducted by women in domestic settings.[39] From the Dublin Society's *Proceedings*, it is possible to determine that Elizabeth Cortez of Co. Cork, Elizabeth Gregg of Co. Clare, Mrs Campbell and Martha Charlotte Menzies of Dublin and Miss J. Fitzgerald were all involved in rearing silkworms between 1765 and

1804. On 19 December 1765, Cortez sent the Dublin Society 'A considerable Quantity of Cocons and raw Silk' from her home in Inishannon.[40] The Society had the specimen examined by a dealer, who declared the cocoons 'fuller than usual, and the Silk perfectly good in it's [*sic*] Kind, and worth *1 l.* 3 *s.* per lb'.[41] Although the original letters do not survive, Cortez's correspondence with the Society lasted several years and her efforts were rewarded with sums of money on at least three occasions.[42] A Huguenot who settled in Inishannon, Cortez formed part of a group of refugees who were helped to establish businesses by La Société pour les Protestants Réfugiez. Support of the Inishannon textile manufactory, for which Cortez's raw silk was intended, was an important focus of this initiative.[43] On 10 March 1774, it was reported that she had also furnished the Society with a copy of her 'Memorial of her Observations on the breeding of Silk Worms'.[44]

Other women were less successful in securing a cash reward for their work. Elizabeth Gregg, also writing in the 1760s, tried several approaches to gain the notice of the Dublin Society. In 1768, she sent 'A considerable Quantity of raw Silk produced by Silk Worms in the County of Clare', which was judged 'perfectly good in its Kind' but a proposed bounty of twenty guineas was ultimately rejected.[45] Unperturbed, Gregg's work came in front of the Society less than a year later when a wealthy patroness, Lady Anne O'Brien, presented 'a Piece of *Irish* flowered Silk' made from Gregg's raw product.[46] Unfortunately, despite the bounty of twenty guineas being re-proposed and the decision twice postponed, it was still rejected on 10 May 1770 by a majority of two-thirds.[47] Similarly, in 1804, all Miss J. Fitzgerald could secure for her specimen of raw silk 'equal [in quality] to any imported' was 'the Society's thanks for her patriotic exertions'.[48]

Two women based in Dublin achieved more than Gregg and Fitzgerald both in the scale of their operations and the recognition they were able to extract from the Society: Mrs Campbell and Martha Charlotte Menzies, both writing in the first years of the nineteenth century. Campbell worked on this project with her daughters from 45 Charlemont Street and on 9 July 1801 she applied for funds to 'purchase machines and procure hands to assist, by which means she has no doubt of becoming a complete silk glove manufacturer, and be able to sell silk gloves much cheaper and better than they can at present be bought in *Dublin*'.[49] Similarly, Menzies of Pembroke Quay in Dublin had studied 'with the strictest attention and perseverence' both 'the feeding and manageing [*sic*] of Silk-worms, and the proper method of winding and preparing the silk for the weavers' use'.[50] However, she needed the Society's help to 'enable her to carry on the raising of the worms in an extensive manner', specifically by affording to take on two or more child apprentices.[51] Both women secured significant contributions to their enterprises: Campbell ten guineas in 1801 and a further five in 1802 and Menzies twenty guineas in 1802. The emphasis they placed on the need to teach others the techniques they had learned was echoed in the earlier submissions of Elizabeth Cortez who had, likewise, stated her willingness 'to instruct [a] young Person in the Management of Silk Worms, and the Art of winding Silk after the Manner practised in France' and 'taught her Art of Raising Silk Worms to Mrs Anna Bell, who was now so well experienced therein, as to be able to instruct others'.[52]

It is difficult to trace the social status of all of the women mentioned in the Dublin Society's *Proceedings*, but it is clear that Martha Charlotte Menzies was a member of the landed gentry.[53] Elizabeth Cortez was part of the Huguenot community and while French Calvinism had found believers amongst all ranks of society, this form of Protestantism was particularly prevalent amongst literate craftspeople, hence the Huguenots' reputation for bringing technical expertise to Ireland and Britain. It is likely, therefore, that Cortez hailed from this section of society. It also seems significant that both Cortez and Elizabeth Gregg sought the support of members of the landed class in putting their case to the Society – Gregg gained the support of the wealthy O'Brien, and the MP and landowner, Thomas Adderley, was enlisted to present Cortez's work, suggesting both of these women stood to gain from this elite endorsement.[54] This practice of lower-status individuals securing a guarantor for the quality of their work is also recognisable in the records of the Society for the Encouragement of Arts, Manufactures and Commerce.[55]

In the English sources, two avid silkworm experimenters stand out: Mrs Ann Williams who lived in Gravesend in Kent and Miss Henrietta Rhodes (1756–1817/8) of Bridgnorth in Shropshire.[56] Both women submitted many letters to the London Society describing their home experiments in cultivating silkworms, which offer detailed descriptions of their activities and motivations – these form the focus of the analysis below.[57] At the time of her writing, Williams was the postmistress at the Gravesend Post Office; she lived alone – her father having died several years earlier – but she made use of at least one servant.[58] The manner in which Williams conducted her experiments suggests that her domestic environment was relatively confined as she adapted key living space to serve experimental ends. Prior to embarking on this venture, Williams had published a volume of her own poetry, which she dedicated to the Postmaster General, H. F. Thynne (later Lord Carteret).[59]

Henrietta Rhodes lived at Cann Hall, a sizeable mansion dating back to the sixteenth century most likely occupied by the lower gentry.[60] She never married and, over her lifetime, undertook a number of literary and intellectual pursuits, publishing several works.[61] Henrietta Rhodes referred more frequently to re-deploying domestic servants to her silkworm project, suggesting greater access to this resource, although she conducted a large amount of the work herself.

Whilst Williams was concretely middling sort, with a job to hold down, Rhodes was a leisured gentlewoman with considerable time and space to put to the task of rearing silkworms. As *The annual biography and obituary* informed its readers, 'although never successfully wooed herself, yet she wooed the muses'.[62] In this volume and an obituary in the *Gentleman's Magazine*, Rhodes was acknowledged as having published three works, including poems and essays, a novel and an account of Stonehenge. Both obituaries were agnostic on the quality of her literary output, suggesting that whilst she 'possessed a comprehensive mind' her novel divided public opinion and her poetry 'did not rise above mediocrity'.[63] That said, her eighty-page *Poems and miscellaneous essays* (1814) was funded by subscription, revealing an extensive network, including 'many of the first nobility and gentry of the land' and, as the *Gentleman's Magazine* observed, 'such a profusion of illustrious names is rarely to be seen, being principally obtained through the interest and connexions of a few particular friends in the higher circles, who were much devoted to her welfare'.[64]

The societies in London and Dublin might have provided an impetus for women to develop their domestic investigations and

– importantly for historians – report upon them, but these observers and experimenters were rooted in their local contexts and this institutional apparatus only provided one dimension to their endeavours. As David Livingstone's seminal work has shown us, 'Each site provides repertoires of meaning' and as social and material interactions shape discourse, so 'scientific knowledge bears the imprint of its location'.[65]

Home experiments

In 1778, the Society for the Encouragement of Arts, Manufactures and Commerce gave a bounty of twenty guineas to Mrs Ann Williams and, seven years later, a silver medal to Miss Henrietta Rhodes for her efforts in rearing silkworms and producing silk. Both women were engaged with literature concerning the production of silk and discussed this project with others, and Rhodes referred directly to Williams's earlier submissions to the Society, revealing that she was also a reader of its *Transactions*.

Establishing a colony

On 14 October 1777, Williams sent her first letter on this topic, reporting that she had 'forty-seven Silk worms spinning, which were but one month old yesterday; the first span on Friday last, and are in fine cocoon; those of Saturday, Sunday, and yesterday, are forming them'.[66] Whilst Williams thought she could have hatched more, she declared this a sufficient 'specimen of what may be done by a watchful attendance and industry'.[67] She also offered to 'send my Silk up next week by a friend, under three different classes, that of my first brood, that of my second, and some reeled off the eighteenth, nineteenth, and twentieth of November' for inspection by Society officials.[68] Similarly, when Henrietta Rhodes began to write to the Society in 1784 she provided details of her practice, which had started in the summer of 1782 when a friend sent her 'a dozen and half of Silk-worms'.[69] Rhodes admitted to being, at this time, 'totally ignorant of the method of treating them', but by the following May of 1783:[70]

> I found my stock increased to about thirteen hundred, and I was so fortunate as to lose very few during the whole time of feeding; for I had twelve hundred and seventy very fine Cones, and they produced me near four ounces of silk. I preserved all the eggs from these; and on the 12th of last May, placed them in the sun: they were hatched in incredible numbers; and, by the most accurate calculation, I was mistress of more than ten thousand.[71]

The naturalist and entomologist, René Antoine Ferchault de Réaumur, thought it crucial to have a great number of live specimens to watch at home.[72] Whilst the motivation for Rhodes was, partly, the quantity of silk that could be produced, large colonies were considered advantageous from the point of view of natural history. Rhodes also enclosed a sample of her own silk, the evidence required to be considered for a prize, claiming that 'many good judges' had declared it 'superior to any that has, yet been manufactured in England, and equal to that which comes from Italy'.[73] In this way, both women used a combination of reported experiments and samples of their product to provide evidence of their enterprise and prove their eligibility for the Society's notice.

Throughout these letters, the pressing need to secure a reliable supply of food for the worms was apparent. During the several years that Rhodes had been rearing silkworms, her colonies had grown rapidly and she was forced to harvest mulberry tree leaves in a ten-mile radius of her home and employ the help of friends to secure sufficient quantities: 'I sought after Mulberry-trees with an anxiety I cannot describe, and the discovery of a new one was a real acquisition.'[74] Ann Williams tested a number of locally available options, including lettuces and blackberry leaves, as 'food for my little family', finding the latter 'they eat surprisingly, and grew amazingly'.[75] However, Williams's 'researches ... did not stop here'; she 'Next presented them with the young and tender leaves of the Elm, which they devoured with great avidity. Cowslip leaves, and flowers, they are very fond of'.[76] Once Williams was able to procure mulberry leaves, she found that her worms 'would not touch' any of these other foodstuffs.[77]

Later, Henrietta Rhodes would criticise Williams's approach, commenting that 'Mrs. Williams's observations on the various kinds of leaves they will eat, admitting their truth, can never be of the least utility, unless to gratify the curiosity of the speculative philosopher.'[78] Williams was certainly more concerned with observing the processes she put in motion than making her activities immediately profitable, identifying herself as one of 'those who love to pry into the secrets of nature'.[79] To this end, Williams pondered the reasons for the worms' preferences, writing on 19 October 1777, 'It is worthy [of] remark, they will not touch a red flower; ... and they seemed to avoid them with a kind of horror. I suppose nature debars their feeding on them, as it might hurt the colour of the silk.'[80] Whereas Williams reflected broadly on the workings of nature, Rhodes was more practically focused – trying 'most of the different leaves to be found in a large kitchen garden' when a 'scarcity of food ... threatened me' but remaining intent on procuring 'sufficient quantities to serve a manufactory'.[81]

Domestic material culture

Both Williams and Rhodes adapted spaces in and around their homes to cultivate their colonies. In her second letter, written just five days after the first, Williams described the conditions in which she kept her worms and the attention she paid them on a daily basis:

> I keep them in a woman's large hat box, feed them every day at Ten o'clock; at Four in the afternoon, and Eleven at night; keeping them very clean. When I clean them I remove them as follows: In a Morning they are always upon the leaves, I take them out gently upon them, and when the box is cleaned, I lay them in, on the same leaves, with fresh ones over them (with the dew on, if I can get them) and the fibre side of the leaves up: when they are all on the upper leaves, I remove the old ones; by this method a quantity of silk is saved.[82]

Imaginative re-purposing of existing household objects allowed the enterprise to fit neatly into the spaces offered by her home environment, which no doubt formed part of the Post Office premises she ran during this time. When Williams was concerned about the temperature of her brood affecting their hatching, she 'put the papers with the Eggs, into a pigeon-hole in a Cabinet, nearly opposite the fire. As soon as the frost set in, I covered the hole with paper several times double, to keep out the night air.'[83] Furniture of everyday use was promptly re-purposed as a home for silkworms, as the need arose, revealing that investigators like Ann Williams found the tools and affordances she required amongst the material culture of her domestic space. In pressing these objects into the service of science, Williams's colony came to nestle at the heart of her household, next to the fireside.

Unlike Williams, Rhodes cultivated her silkworms in a space specially designed, which she referred to as a 'manufactory'. Living in this place, Rhodes's silkworms were 'so situated that they were exposed to all the sounds incidental to a country town, from the barking of dogs, up to a family concert; and I am sure they never were visibly affected by either'.[84] So, whilst this was a space external to the main house, the manufactory was certainly near to home, most likely a domestic outbuilding, and situated close to her local community in Bridgnorth. However, like Mrs Campbell of Dublin whose enterprise outgrew her household, leaving her with 'the necessity of destroying multitudes of these valuable creatures for want of room', it seems that Rhodes's earliest endeavours had been conducted in the main house.[85] Her own home offered sanctuary once again when disaster struck her colony during an unseasonal cold snap:

> It was sufficiently obvious that the making of fires would remedy the evil; but they were unfortunately situated over a range of warehouses, which rendered that, not only dangerous, but impossible. To remove such numbers [of worms] into the house, was equally impracticable; but alas! They were soon sufficiently reduced for me to adopt that plan, and in one of the coldest days I almost ever felt, with the assistance of several of my friends, I removed them to their former apartment. Here I kept large and constant fires, and the Worms as they arrived at maturity, pursued their industrious occupations with alacrity.[86]

When faced with calamity, it was the technology of home that could rescue the situation as the fireside, once again, proved the place most likely to preserve the remaining silkworms and keep them spinning. With reference to a manufactory, a 'former apartment' within the main house and a kitchen garden in which emergency food could be found, the household space open to Rhodes was more extensive than that of Williams and flexible in terms of its use. This description certainly accords with the evidence that Rhodes lived at Cann

Hall, an old and substantial home considerably remodelled in the nineteenth century.

Regimes of labour

Henrietta Rhodes made a point of relaying details of how her work with the silkworm manufactory fitted into her day. Rhodes 'fed them three times a day with leaves which had been gathered in the morning' and once a week 'the pans were to be cleaned' and 'in that office I was assisted by a servant'.[87] She was happy to report that this regime was not so onerous that it kept her from 'other avocations' or 'amusement'.[88] Indeed, obituaries of Rhodes referred to the full intellectual life she had led, taking on her own writing projects and contributing to those of others.[89] Williams, on the other hand, found herself more pressed for time, complaining in earlier letters to the Society that her Post Office duties kept her from preferred pursuits, comparing the work to 'Egyptian Slavery', offering 'no rest night or day'.[90] Nevertheless, by the time she embarked on her silkworm project, Williams was able to feed her worms three times a day (at 10 am, 4 pm and 11 pm), collect the leaves upon which the worms sat (first thing every morning, preferably with the dew still on them) and use these to replace the existing leaves.[91] In addition, there were periods when the silk required collecting and measuring and Williams also spent time observing the activity of the worms.

The routines outlined in these letters hint at the rationale for cultivating silkworms in domestic environments and for the role of women in this cottage industry. The worms required attention for short bursts of time at fairly regular intervals and multiple times a day. They thrived in warm, dry environments – easily accessible to the cultivator – that were common in a well-heated household. Dubliner Martha Charlotte Menzies also considered the work of silkworm rearing to be compatible with other domestic work when she planned to take on apprentices, committing to 'instruct not only in the silk business in its season, but also in all kind of domestic and useful work, which would give them the means of obtaining by industry, a comfortable living'.[92] Silkworm rearing and silk harvesting were thus easily accommodated by the rhythms and routines of domestic labour. However, to maximise productivity and satisfy curiosity, silkworm breeders were drawn into practices of close observation and experimentation – concerning themselves with the interplay of the insects' characteristics and the particularities of domestic and neighbourhood environments.

An American venture

Since the early seventeenth century, great hopes had been invested in American soil for the cultivation of silkworms and the mulberry bushes upon which they fed. King James I (and VI of Scotland) even sent bushes and silkworm eggs directly to the colonies for that purpose. By the eighteenth century, there was frequent communication between writers in England, Ireland and America about this enterprise. For example, in April 1756, an anonymous correspondent wrote to the *Gentleman's Magazine*, noting the interest of both the London and Dublin societies in the subject and the particular suitability of America for the raising of silkworms.[93] Commenting on how silk production had gifted the 'vast riches of *China*' and the 'extraordinary treasure for the king of *Sardinia*', the author remarked upon the similarities in climate between Pennsylvania, Maryland, North Carolina, Virginia and Georgia and 'Nanking', China.[94] The article emphasised the advantages of cheap land, 'a great number of hands' ready for the task (including enslaved peoples), the profitability of the enterprise and the potential for a range of grades of silk to be produced in the varying climate and conditions of the southern Atlantic seaboard, which produced two different species of mulberry bush and, possibly, different kinds of silkworms.[95]

In 1770, Samuel Pullein's guidance was re-printed in Philadelphia by the Quaker printers, Joseph Crukshank and Isaac Collins, with a preface 'giving some account of the rise and progress of the scheme for encouraging the culture of silk, in Pennsylvania, and the adjacent colonies'.[96] Moreover, the Preface to volume one of *Transactions of the American Philosophical Society* (1769–71) noted Pennsylvania's climate and conditions as particularly promising in respect of the cultivation of a silk industry.[97] The American Philosophical Society has been established in 1743, twelve years after the founding of the Dublin Society and eleven years prior to the London-based Society for the Encouragement of Arts, Manufactures and Commerce.

Like these other societies, the raising of silk was a topic of concerted interest, with a 'Silk Society' initiated under the auspices of the Committee on Husbandry and American Improvements.[98] These eighteenth-century efforts did indeed result in silk manufacture becoming established in Pennsylvania by the early nineteenth century.

In the first volume of the American *Transactions*, descriptions of home experiments with raising silkworms were printed and their author was Moses Bartram. Bartram (1732–1809) was the second son of the well-known Anglo-American botanist and horticulturalist John Bartram (1699–1777). As such, he grew up among his father's botanical gardens and, alongside his brother Isaac, became an apothecary and man of considerable social standing in the city of Philadelphia.[99] In 1766, Moses Bartram was elected to the American Society for Promoting Useful Knowledge and he had wide-ranging interests including the effect of lakes on the climate, sleep-walking and locusts.[100] However, his observations on silkworm breeding act as a helpful comparison with Williams and Rhodes and underline that the home was a space of experimentation for men as well as women.

As the Preface to the first volume of the American Philosophical Society's *Transactions* had revealed, there were some particularities to silkworms in Pennsylvania. First, in 'this part of America, different kinds of Silkworms are found upon different trees and shrubs'.[101] Not all the American silkworms fed primarily on mulberry, and 'those that feed on the Sassafras, are larger, and the Silk they produce, though not so fine, is much stronger than that of the Italian Silkworm'.[102] Moses Bartram's account begins with a personal encounter with these 'wild silk worms' and his silkworm-collecting 'excursion along the banks of Schuylkill' yielded a 'lucky' five cocoons with live 'nymphæ' within.[103]

When Bartram got his five cocoons home, he placed them 'in my garret opposite to a window, that fronted the sun rising', explaining that 'the warmth of the sun might forward their coming out'.[104] Unfortunately, when the first fly emerged 'it made its escape', the window having been left open.[105] A week later, another fly hatched 'out of a large loose pod' and 'began to lay eggs', the next two – males – 'grew very weak and feeble and unable to fly' and within two days they had both died. After a full week of laying 'near three hundred eggs' the female fly also expired. Six days later, the fifth and last pod produced 'a large female fly, of the brown kind like the rest'.[106] Like the first female, this fly laid hundreds of eggs but Bartram doubted the likelihood of success, given the absence of a male to fertilise. In the end, both the eggs from the first and second female 'began to shrivel and be indented in the middle' and failed to produce live offspring. Nonetheless, Bartram 'folded them all up in separate papers and laid them by, to see if any would hatch the spring following'.[107] Disappointingly, the following year the eggs remained bereft of worms 'from whence' Bartram 'concluded they had not been impregnated by the males'.[108]

Despite the manifold failures of this first attempt, Bartram was 'determined to make another trial' but this time 'with more caution and circumspection'.[109] He foraged further afield for the cocoons, gathering specimens from swamps and uplands and from several different kinds of trees.[110] However, he persisted with the positioning of them in the sunny spot by his garret window. This time, he took no chances with escapees and 'tacked course cloths up against the windows on the inside', which also prevented them from damaging themselves in colliding with the glass, which Bartram thought might have 'prevented their copulating'.[111] His efforts were not in vain; this time no flies escaped and between late May and June fertilised eggs hatched and produced worms. The abundant offspring gave Bartram the challenge of sourcing food and, similarly to Williams, he tried out several kinds of leaves and vegetables – the worms settling on alder as a preferred meal despite the availability of mulberry leaves.[112] The experimentation with food was not without its losses, with several being killed in the process of 'shifting them from one kind of food to another'.[113] Later, Bartram reflected 'From sundry experiments, I found the worms averse to changing their food. On whatever they first begin to feed, they keep to it.'[114] However, many of Bartram's worms stopped feeding altogether, 'shrunk up short, and seemed motionless', which caused him to worry that 'all my hopes of raising them were frustrated' and conclude that 'they would perish'.[115]

However, this second trial was a success and Bartram was 'agreeably surprized [*sic*] to see the little animals ... creeping out of their old skins, and appearing much larger and more beautiful than before'.[116] Bartram's account takes on the naturalist's concerns with detailed physical description, including the 'beautifully spotted' appearance of the large brown fly that emerged from one of his 'pods' and the observation that male flies were smaller but with much brighter 'and more beautiful' colours.[117] Bartram used anthropomorphic language, commenting that the worms 'devouring their old coat ... seemed a delicious repast to them'.[118] His descriptions of their appearance acted similarly: 'It is remarkable every change they undergo adds fresh beauty to the worms, and in every new dress, they appear with more gaudy colours and lively streaks.'[119] Having conducted one fruitful trial, Bartram remarked that, despite the worms being hatched within three days of one another, the interval between the first and last worm commencing spinning was 'no less than nineteen days'.[120] Using empathetic language, he wondered whether 'this was owing to the weakness or strength of the vital principle in some more than others', the switching from one food to another, or to 'their being frightened, and thereby prevented from feeding'. He concluded that 'Farther experiments may possibly explain the matter.'[121]

Despite the successes of Bartram's efforts, disaster struck on 20 June, when one worm 'was destroyed by a kind of bug armed with a long bill with which it pierced the side of the worm, and sucked out its vitals'.[122] In general, Bartram's account emphasises the particular suitability of his region for breeding silkworms and argues that only the native species should be reared on account of their amenability to this local environment and its available foodstuffs. However, this episode also shows that there were also well-adapted predators which posed a threat to the local silkworm varieties:

> Its bill is so long, that it can stand at some distance from the worm, and with its weapon wound it, notwithstanding the bunches of hair or bristles, in form of a pencil, with which the worm is covered, and which are its principal defence.[123]

In this instance, Bartram had made the mistake of bringing in the bug with the leaves that would feed the worms. However, the rest of his account focuses in some detail on the ways in which he augmented the silkworms' space in order to keep them safe and encourage the all-important task of spinning.

Much like Ann Williams, Moses Bartram used materials ready to hand in his domestic space to improve the colony's living conditions and to aid its productivity. Initially, when his worms were 'in search of a proper place to spin', Bartram 'got sticks, in which I fixed a number of pegs for the greater conveniency of the worms', although he noted that 'they can spin in any place, where they can form an angle for their webs'. Indeed, one of Bartram's number did use the angle of the corner of the garret window to do so.[124] As the silkworms were inclined to wander about a great deal before deciding on a place to spin, Bartram took his construction of sticks in the form of a rack and fixed them 'in glass bottles to prevent the worms from getting off'.[125] Bartram's homemade solutions also extended to feeding the silkworms, which required the leaves of trees that did not grow in abundance in the city of Philadelphia where he lived. He arrived at a 'method … with the least trouble to myself' whereby:

> I filled several bottles with water; in these bottles I placed branches of such vegetables as the worms feed on. I placed the bottles so near each other, that when any of their food withered, the worms might crawl to what was fresh. By this means I kept their food fresh for near a week.[126]

Finally, Bartram proposed a more elaborate system of narrow troughs, with notched edges, to allow 'pieces of straight wood [to] be fixed, so that the branches, on which the worms are to feed, may lie in the notches, and their ends be fixed under the piece of wood at the bottom'.[127] The formation offered the silkworms direct access to food and water and the caretaker the ability to easily refresh the water supply so that it remained 'sweet and clean'.[128] This more elaborate apparatus further developed the improvised version that Bartram had devised iteratively over the course of his initial 'experiments'.[129]

In addition to the troughs and racks for cultivating silkworms, another box design was advised for the process of extracting the silk from the cocoons. Again, the boxes were constructed of common materials, 'strips of wood' and nails, and 'washed with a solution of gum Arabic, or cherry tree gum'.[130] Whilst gum Arabic was not necessarily a domestic essential, its binding properties made it useful in the making of some common household products including iron-gall inks and it was also put to use in the sizing and dyeing of silk, cotton and other textiles.[131] Bartram advised his reader in this prescriptive way because he was 'persuaded' that silkworms 'might be raised to advantage, and perhaps, in time, become no contemptible branch of commerce'.[132] He boasted of his robust, native silkworms that their seasonal schedule of hatching and spinning ensured that 'they are not subject to be hurt by the frost' and that 'they lie so long in their chrysalis state, the cocoons may be unwinded at leisure hours in the ensuing winter'. Unlike 'foreign worms', 'Neither lightnings nor thunder disturb them', and the cocoons were at least four times larger than their counterparts overseas, thereby offering a greater yield of silk.[133] Thus, with a method developed through close observation, experiment and the everyday materials and spaces of home, Moses Bartram offered those 'who have leisure' the encouragement 'to make further trials'.[134] His hopes for this enterprise included a profitable industry in his home county in Pennsylvania, but he also recognised the potential to put idle or leisured hands to work in a task that could fit, neatly, into a sunny attic corner.

Conclusion

Whilst British and Irish silkworm breeding advocates emphasised the leisured, the idle and the female in their characterisation of the workforce for this new industry, the examples discussed here also include men and working people.[135] Ann Williams offers a striking example of a working woman with relatively limited domestic space engaging fully with this complex endeavour and being formally rewarded for her efforts. In other parts of the world, where the practice was long-established, silk production was associated with domestic space and this, no doubt, shaped the ideas of British, Irish and American promoters of the industry. Those who wrote about their own home experiments and sent samples of their work to formal institutions often emphasised the ease with which this pursuit fitted into the home and its existing regimes of labour.

However, as the more detailed descriptions discussed in this chapter show, the precise spaces and approaches varied considerably. In some cases, peripheral areas of the household, such as outhouses or garrets, offered accommodation for these colonies. Elsewhere, silkworms were kept close to the hearth. Tensions existed between modes of writing that emphasised a scalable and profitable industry of interest to a national society and approaches that aimed to share the natural history observations that were key to keeping silkworms alive and producing the greatest amount of silk. The testimony of Ann Williams, Henrietta Rhodes and Moses Bartram offers a combination of both, although Rhodes was especially concerned with maintaining a business-oriented narrative.

Clearly, synergies existed between the temporal rhythms of domestic labour and the schedules of feeds required by silkworms. On a larger scale, the seasonal changes felt by households, in terms of heating and provisioning, were mirrored by the cycles of silkworm reproduction and spinning. Rather than seeing this activity as something that fitted neatly with an existing domestic regime, one might just as easily see it as emerging from that regime. The individuals discussed here, most especially women with responsibility for the labour of home (by hand or order), were well positioned to experiment with silkworm rearing, in terms of their skills, schedules and command of household space and material culture. The multifaceted nature of silkworm breeding, in terms of its relationship with natural history, manufacture and commerce, is revealing of the interrelated nature of these different pursuits, especially as they came to pass within the context of the home.

Notes

1 Dear, 'Meanings of experience', p. 106; Latin terms: *experientia* and *experimentum*.

2 Dear, 'Meanings of experience', p. 115; Étienne Chauvin quoted from *Lexicon philosophicum*, vol. 2 (Leeuwarden, 1713; facsimile repr. Düsseldorf: Stern-Verlag Janssen, 1967), p. 229. According to Chauvin, this experience came in several forms: accrued through everyday life, from focused examination and for the explicit purpose of uncovering truth.

3 Francis Bacon's (1561–1626) writings were key to all of these developments; see for example Antonio Pérez-Ramos, *Francis Bacon's idea of science and the maker's knowledge tradition* (Oxford: Clarendon Press, 1988); Stephen Gaukroger, *Francis Bacon and the transformation of early-modern philosophy* (Cambridge: Cambridge University Press, 2001); John Henry, *Knowledge is power: Francis Bacon and the method of science* (Cambridge: Icon Books, 2002).

4 Terrall, *Catching nature*, p. 51.
5 *Transactions of the Society, instituted at London, for the encouragement of arts, manufactures, and commerce*, vol. 14 (1796), pp. 295–98.
6 *Ibid.*, p. 295.
7 *Ibid.*, pp. 295–6.
8 *Ibid.*, p. 297.
9 *Ibid.*
10 McCann, 'Private laboratory'.
11 As McCann highlights, William Constable of Humberside had a Philosopher's Room set up in his house in 1769 to house his collection of scientific instruments. Likewise, John Stuart, the third Earl of Buter, amassed significant quantities of scientific instruments; 'Private laboratory', p. 635; see also Werrett, *Thrifty science*.
12 McCann, 'Private laboratory', p. 637.
13 This painting is still at Petworth House.
14 For more on landowners' developing interests in this sphere, see James Fisher, 'The master should know more: Book-farming and the conflict over agricultural knowledge', *Cultural and Social History*, 15:3 (2018), pp. 315–31.
15 McCann, 'Private laboratory', pp. 637–8.
16 *Ibid.*, pp. 639–40, 642.
17 This includes over eighty pieces of glassware; there is also a small amount of earthenware – four retorts, an alembic and three crucibles – although it is possible that these artefacts were used in the dispensary instead of the laboratory; see McCann, 'Private laboratory', pp. 640–55.
18 See Joan Thirsk, *Economic policy and projects: The development of a consumer society in early modern England* (Oxford: Clarendon Press, 1978), pp. 7, 120–2, 130; and Linda Levy Peck, *Consuming splendor: Society and culture in seventeenth-century England* (Cambridge: Cambridge University Press, 2005), pp. 1, 14, 16, 31, 73, 85–92, 106–10.
19 Levy Peck, *Consuming splendor*, pp. 89, 93, 99–103; Ann Leighton, *American gardens in the eighteenth century: 'For use or for delight'* (Boston, MA: Houghton Mifflin, 1986), p. 233.
20 *Transactions of the Society*, vol. 11 (1793), pp. 220–3; Marten himself was engaged in another tree-growing project – that of cinnamon trees – and he trialled alternate planting of mulberry bushes and cinnamon trees in the new plantations of Madras, in order to provide the necessary shade for the valuable spice tree to prosper. This innovation was developed after Mrs L. Anstey (née Light) first successfully planted cinnamon trees in this region; see p. 212.

21 For more detail on Irish silk manufacture, which suffered from the British government's policies on imports, see Mairead Dunlevy, *Pomp and poverty: A history of silk in Ireland* (London: Yale University Press, 2011), pp. 29–60.

22 Ireland and Scotland led the way with initiatives of this kind, as there was a short-lived Honourable Society of Improvers in the Knowledge of Agriculture established in Edinburgh (1723–45), the Dublin Society was founded in 1731 (becoming the Royal Dublin Society after 1820) and the Society for the Encouragement of Arts, Manufactures and Commerce was founded in London in 1754, acquiring its Royal charter in 1847.

23 For more on the way the Society sought to embody an encompassing 'public', see Paskins, 'Sentimental industry', p. 27.

24 Clark, *Sociability and urbanity*, pp. 2–4, 57–8, 274–6; Celina Fox, *The arts of industry in the age of enlightenment* (New Haven, CT: Yale University Press, 2009), pp. 179–82.

25 *Transactions of the Society*, vol. 2 (1784), p. 153, included in the 'Summary account of rewards bestowed by the Society' 1775–82; likewise the Dublin Society offered Anthony Crouset a £100 interest-free loan for 'raising white Mulberry Trees' on 15 January 1761; see Dublin Society Minute Book, vol. 6 (9 Mar. 1758–13 Aug. 1761).

26 Royal Society of Arts (hereafter RSA), PR.GE/112/13/5, p. 18.

27 Calls were issued in 1768, 1769, 1776, 1783, 1784, 1786, 1787, 1788 and 1789, demand intensifying around the time that Henrietta Rhodes wrote to the Society (1785–6), see RSA, PR.GE/112/13/5; PR.GE/112/13/6; and PR.GE/112/13/7.

28 Samuel Pullein, *Some hints intended to promote the culture of silk-worms in Ireland* (Dublin, 1750); Christian Schultze, 'A memoir on the great advantage of raising silk-worms and cultivating bees in Ireland', *Dublin Society transactions*, vol. 1, pt. 2 (1799), pp. 75–88.

29 Pullein, *Some hints*, pp. 12, 15; Pullein also sent a letter on the culture of silk to the Society for the Encouragement of Arts, Manufactures and Commerce, in London, on 9 Dec. 1758; see RSA, PR/GE/110/3/23 and published pieces on 'A new improved silk-reel' and 'An account of a particular species of cocoon, or silk-pod, from America' in the Royal Society's *Philosophical transactions*, vol. 51 (1759–60), pp. 21–30 and 54–7 respectively.

30 *The Dublin Society's weekly observations*, 1, no. 1: 4 Jan. 1736–7 (Dublin, 1739), p. 7.

31 *Ibid.*, this approach broke with the discourse of 'secret' knowledge that had dominated cultures of knowledge in previous centuries; see Elaine Leong and Alisha Rankin (eds), *Secrets and knowledge in medicine*

and science, 1500–1800 (Farnham: Ashgate, 2011), but had something in common with centuries-old efforts by governing circles to promote practical projects that could effectively exploit material things for the betterment of society, see Thirsk, *Economic policy*.

32 Anton Howes, *Arts and minds: How the Royal Society of Arts changed a nation* (Oxford: Princeton University Press, 2020), pp. 1–28. The Society for the Encouragement of Arts, Manufactures and Commerce's founding members comprised nobility, gentry, clergy and merchants (including four Fellows of the Royal Society), the Dublin Society was founded by fourteen Anglo-Irish Dubliners, including medical men, two clergymen and a landowning lawyer. See Fox, *Arts of industry*, pp. 182, 186; and James Meenan and Desmond Clarke (eds), *The Royal Dublin Society, 1731–1981* (Dublin: Gill and Macmillan Ltd, 1981), pp. 1–3.

33 Fox, *Arts of industry*, p. 187.

34 Meenan and Clarke, *Royal Dublin Society*, p. 5.

35 Correspondence with Anton Howes (historian in residence at the RSA) revealed that, in his view, the class make-up of people writing to the Society for the Encouragement of Arts, Manufactures and Commerce varied across category, noting 'a very high proportion of gentry and even nobility' in agriculture, but an 'overwhelming majority' of manufacturers or merchants in the fields of manufactures, mechanics and colonies and trade.

36 Again, I am grateful to Howes for this information, which formed part of his research for *Arts and minds* and this percentage refers to all premiums, bounties and thanks given by the Society to individuals for their submissions. It has not been possible to ascertain comparable data for the Dublin Society during this period.

37 Fox, *Arts of industry*, pp. 183, 187. From 1772 the Postmaster General committed to disseminating free of charge copies of the lists of premiums to all post offices in Great Britain, Ireland and America; see Fox, *Arts of industry*, p. 191. From 1736 onwards the Dublin Society published a weekly paper on an aspect of their work in the Dublin Newsletter, disseminating their findings widely; see Meenan and Clarke, *Royal Dublin Society*, p. 5.

38 Paskins, 'Sentimental industry', pp. 118–19.

39 Pullein, *Some hints*, p. 16.

40 *Proceedings of the Dublin Society*, vols 1–2 (15 Mar. 1764–2 Oct. 1766), pp. 261–2.

41 *Ibid.*

42 In 1765: 22 l. 3 s.; 1766: 22 l. 15 s. and 3 l. 18 s.

43 See Grace Lawless Lee, *The Huguenot settlements in Ireland* (Berwyn Heights, MD: Heritage Books, 2008), pp. 83–5; Cortez may have

arrived in Ireland 1752 as a result of efforts in the Languedoc to enforce Catholic baptism on Protestants.

44 *Proceedings of the Dublin Society*, vol. 10 (Oct. 1773–Aug. 1774), p. 367.

45 *Proceedings of the Dublin Society*, vol. 5 (Oct. 1768–Jul. 1769), pp. 261, 285.

46 *Proceedings of the Dublin Society*, vol. 6 (Oct. 1769–Aug. 1770), p. 39; Anne O'Brien was the wife of Sir Lucius O'Brien of Dromoland Castle, Co. Clare, who was a politician and member of the Dublin Society. Anne was a named patroness amongst a group of well-to-do women called 'Encouragers of the Irish silk Ware-house'.

47 *Proceedings of the Dublin Society*, vol. 6 (Oct. 1769–Aug. 1770), p. 183.

48 *Proceedings of the Dublin Society*, vol. 41 (1 Nov. 1804–15 Aug. 1805), pp. 5, 9–10.

49 *Proceedings of the Dublin Society*, vol. 37 (6 Nov. 1800–30 Jul. 1801), p. 170.

50 *Proceedings of the Dublin Society*, vol. 39 (4 Nov. 1802–11 Aug. 1803), p. 2.

51 *Ibid.*

52 *Proceedings of the Dublin Society*, vols 1–2 (15 Mar. 1764–2 Oct. 1766), pp. 261–2; vol. 12 (Nov. 1775–Jun. 1776), p. 3.

53 See the entry on Menzies in *Burke's landed gentry*, vol. 2 (London, 1847), p. 920: she was the daughter of John-Ryves Nettles (d.1785) of Toureen [Tourin], Co. Waterford and Bearforest, Mallow, Co. Cork. Her three brothers were army officers and she married Captain Menzies, of the 62nd regt., and died without offspring, aged ninety-nine, in July 1837.

54 Thomas Adderley of Inishannon, Co. Cork, who applied on his own behalf to the Society for support for his cultivation of mulberry bushes and plans to develop a silk manufactory. Cortez also provided 'a Certificate of several credible Persons of that Country, that the Silk was produced under her Management at Inishannon', *Proceedings of the Dublin Society*, vols 1–2 (15 Mar. 1764–2 Oct. 1766), p. 261.

55 Fox, *Arts of industry*, pp. 191–2.

56 For additional analysis of these women's letters to the Society for the Encouragement of Arts, Manufactures and Commerce, see Paskins, 'Sentimental industry', pp. 111–18.

57 Williams is referred to as 'Mrs' in the Society's documents and as she does not refer to a husband it is likely that she was a widow, although it is not impossible that this title was used to offer an older woman respect, despite her unmarried status.

58 Williams is recorded as a postmistress in the manuscript *Transactions of the Society*: RSA, PR/GE/118/11/935. In 1775 and 1776, prior to writing to the Society about silkworms, Williams had reported her accidental discovery that cuckoo pint (*Arum maculatum*) could be put to use in dyeing; see RSA, PR/GE/118/8/693–695. Williams received thanks from the Society for this contribution and these letters mention her use of a servant to help her remove the stains caused by cuckoo pint. I am indebted to Anton Howes for this finding.

59 Ann Williams, *Original poems and imitations* (London: 1773). One poem in the volume – entitled 'A serious thought on the death of my father' – describes him as 'a father and a friend' who she 'will adore until my life doth end' indicating a close relationship, p. 133; see Chapter 7 for a discussion of Williams's published poetry.

60 See http://search.shropshirehistory.org.uk/collections/getrecord/CCS_MSA271/ (accessed 5 June 2017); the property was adapted significantly in the nineteenth century and demolished in 1957. The wills of Henrietta Rhodes and her father Nathaniel Rhodes confirm their status as gentry; see The National Archives, Prob 11/1602 and Prob 11/1198.

61 See the *Gentleman's Magazine*, vol. 87, pt. 1 (Apr. 1817), p. 374; *The annual biography and obituary for the year 1818*, vol. 2 (London, 1818), p. 385.

62 *Annual biography*, p. 385.

63 *Gentleman's Magazine*, vol. 87, pt. 1 (Apr. 1817), p. 374; *Annual biography*, p. 385; her novel was titled *Rosalie: Or the castle of Montalabretti*, published in Richmond in 1811.

64 *Gentleman's Magazine*, p. 374.

65 Livingstone, *Putting science in its place*, pp. 6, 13. See also Klein, 'Laboratory challenge'; Larry Stewart, 'Experimental spaces and the knowledge economy', *History of Science*, 45:2 (2007), pp. 155–77; Alix Cooper, 'Homes and households' in Katharine Park and Lorraine Daston (eds), *The Cambridge history of science*, vol. 3 (Cambridge: Cambridge University Press, 2006), pp. 224–37; Golinski, *Science as public culture*; and contributing substantially to wider understanding of science and its practice: Latour, *Science in action* and Bruno Latour, *Pandora's hope: Essays on the reality of science studies* (Cambridge, MA: Harvard University Press, 1999).

66 *Transactions of the Society*, vol. 2 (1784), p. 155.

67 *Ibid.*

68 *Ibid.*, p. 162.

69 *Transactions of the Society*, vol. 4 (1786), p. 149.

70 *Ibid.*
71 *Ibid.*, pp. 149–50.
72 Terrall, *Catching nature*, p. 23.
73 *Transactions of the Society*, vol. 4 (1786), p. 149.
74 *Ibid.*, p. 150.
75 *Transactions of the Society*, vol. 2 (1784), pp. 156, 157.
76 *Ibid.*, p. 157.
77 *Ibid.*
78 *Transactions of the Society*, vol. 4 (1786), p. 164.
79 *Transactions of the Society*, vol. 2 (1784), p. 157.
80 *Ibid.*, p. 158.
81 *Transactions of the Society*, vol. 4 (1786), pp. 156, 164.
82 *Transactions of the Society*, vol. 2 (1784), pp. 158–9.
83 *Ibid.*, p. 156.
84 *Transactions of the Society*, vol. 4 (1786), p. 167.
85 *Proceedings of the Dublin Society*, vol. 38 (5 Nov. 1801–26 Aug. 1802), p. 106.
86 *Transactions of the Society*, vol. 5 (1787), p. 144.
87 *Transactions of the Society*, vol. 4 (1786), p. 153.
88 *Ibid.*
89 In particular, she edited a work by her nephew. See *Annual biography*, p. 385; *Gentleman's Magazine*, vol. 87, pt. 1 (Apr. 1817), p. 374.
90 RSA, PR/GE/118/11/939; see also RSA, PR/GE/118/11/937–948 for further references to the toll Williams's work commitments took on her pursuit of science.
91 *Transactions of the Society*, vol. 2 (1784), p. 158.
92 *Proceedings of the Dublin Society*, vol. 39 (4 Nov. 1802–11 Aug. 1803), p. 2.
93 *Gentleman's Magazine*, vol. 26 (Apr. 1756), pp. 161–3.
94 *Ibid.*, p. 161; 'Nanking' refers to Nanjing, capital of the Jiangsu province in modern China and famous for its silk brocade since the thirteenth century.
95 *Gentleman's Magazine*, vol. 26 (Apr. 1756), pp. 161–3.
96 Abbé Boissier de Sauvages, Joseph Crukshank, Isaac Collins, Jonathan Odell, Asa M. Stackhouse and Samuel Pullein, *Directions for the breeding and management of silk-worms: Extracted from the treatise of the Abbé Boissier de Sauvages, and Pullein* (Philadelphia, PA: Joseph Crukshank and Isaac Collins, 1770). Pullein's piece was printed alongside a treatise on the same subject by the French naturalist and encyclopaedist, Abbé Boissier de Sauvages (1710–95), and the book contains extracts from writers dispersed across London, Philadelphia, Dublin, New Jersey and France.

97 *Transactions of the American Philosophical Society*, vol. 1 (1 Jan. 1769–1 Jan. 1771), pp. iv–vii (pp. i–xix).
98 Zara Anishanslin, 'Unravelling the Silk Society's directions for the breeding and management of silk-worms', *Commonplace: The Journal of American Life*, 14:1 (2013), http://commonplace.online/article/unraveling-silk-society/ (accessed 19 August 2021).
99 Randolph Shipley Klein, 'Moses Bartram (1732–1809)', *Quaker History*, 57:1 (1968), pp. 28–34; the Bartrams were Quakers.
100 Shipley Klein, 'Moses Bartram', p. 30.
101 *Transactions of the American Philosophical Society*, vol. 1 (1 Jan. 1769–1 Jan. 1771), p. vi.
102 *Ibid.*
103 *Ibid.*, p. 224: 'Observations on the native silk worms of north-America, by Mr. Moses Bartram. Read before the Society, March 11, 1768'. Ann Williams also referred to her silkworms as 'Nymphs and Swains' in her letter of 3 May 1778, a swain meaning a male lover, a country lad or a shepherd and a nymph referring to a beautiful maiden who inhabits woods and rivers; see also 'Ye Nymphs and Swains, come join with me. A Pastoral Ode in Praise of Peace' by F. Forrest (published 1760); 'Come ye Nymphs and Swains' was a popular song set to music, a printed version appearing in 1795.
104 *Transactions of the American Philosophical Society*, vol. 1 (1 Jan. 1769–1 Jan. 1771), p. 224.
105 *Ibid.*
106 *Ibid.*, p. 225.
107 *Ibid.*
108 *Ibid.*
109 *Ibid.*
110 *Ibid.*
111 *Ibid.*, pp. 225–6.
112 *Ibid.*, p. 226.
113 *Ibid.*, p. 227.
114 *Ibid.*, p. 229.
115 *Ibid.*, pp. 226–7.
116 *Ibid.*, p. 227.
117 *Ibid.*, p. 225.
118 *Ibid.*, p. 227.
119 *Ibid.*, p. 228.
120 *Ibid.*
121 *Ibid.*, p. 229.
122 *Ibid.*, p. 227.
123 *Ibid.*

124 *Ibid.*, p. 228.
125 *Ibid.*
126 *Ibid.*, p. 229.
127 *Ibid.*
128 *Ibid.*, p. 230.
129 *Ibid.*, p. 229.
130 *Ibid.*, p. 230.
131 Gum Arabic was of sufficient commercial interest that it motivated military action by the British in West Africa in 1758; see James L. A. Webb Jr., 'The mid-eighteenth century gum Arabic trade and the British conquest of Saint-Louis du Sénégal, 1758', *The Journal of Imperial and Commonwealth History*, 25:1 (1997), pp. 37–58.
132 *Transactions of the American Philosophical Society*, vol. 1 (1 Jan. 1769–1 Jan. 1771), p. 230.
133 *Ibid.*
134 *Ibid.*
135 Pullein, *Some hints*, p. 12.

Part III

6

Personal experience and authority

This chapter examines individuals who observed and experimented at home and wrote about these experiences in letters to institutions or friends. As discussed in Chapters 4 and 5, personal experience of natural phenomena formed an important and recognised component of knowledge creation in this period. The qualitative detail offered by these correspondents enables analysis of the way knowledge-makers understood their own practices, contributions and selves. The detailed descriptions of their activities and lives are used to consider the relationship between personal experience and knowledge.

In the 1990s, Steven Shapin's *A social history of truth* heavily influenced understandings of the basis of knowledge-making in seventeenth-century England, placing emphasis on gentlemanly standing and conduct.[1] However, recent scholarship has argued convincingly for a re-evaluation of the authority that can be derived from the knowledge of experience and therefore the kinds of people who can be considered 'scientists'.[2] An inclusive model of knowledge-making is borne out by the evidence that follows. As this chapter will show, eighteenth-century knowledge-makers of a variety of stripes communicated a confidence in their own expertise, regardless of the absence of traditional markers of high intellectual or social standing.

Eighteenth-century society was open to the idea that the skills and understanding of working people could contribute to the nation's collective stock of intelligence. In July 1794, the *Gentleman's Magazine* printed an 'Account of a Method of curing Burns and Scalds By Mr David Cleghorn, Brewer in Edinburgh'. The method had originally been 'Communicated in three Letters' by Cleghorn to the famous surgeon, John Hunter (1728–93).[3] The editor commented that this was 'evidently the production of a plain, sensible, well-informed man, who candidly gives us the result of his experience, and who communicates it to the publick from the most benevolent of motives'.[4] In Cleghorn's second letter, he appealed to 'an eminent physician in Edinburgh', Dr Hay, whose 'liberality of sentiment' would ensure that 'a valuable discovery in the healing art should [not] be disregarded ... merely because it happens to be stumbled upon by a person not of the medical profession'.[5] Considerable magazine space had been allocated to share this information, which was printed in full and ran to four pages. Thus, the medical insights of a brewer were offered to the *Magazine*'s readership in detail and alongside a bid for the value of the experience of a plain and sensible working man.

In taking the perspective of the individual, non-elite investigator, this chapter engages with wider debates about selfhood and affect. Here, the act of enquiry is understood as a commitment that had a strong and sometimes fraught relationship with the sense of self. The decision to enquire could be an emotional one. For the seemingly atypical investigators discussed here, negotiations around issues of social status, role and responsibilities were of crucial importance not only to their ability to pursue scientific activity but also for the value they placed upon that activity both for themselves and wider society.

Knowledge made at home

When knowledge is described as 'know-how' it loses some esteem. Nevertheless, *knowing how* to conduct a range of complex material processes was a prerequisite for running an orderly and productive home. For lower-status people, as much as for the new industrialists of this period, the urge to solve problems and to create new and better ways of doing useful things was a powerful driver of enquiry.

Here, letters written to institutions form the basis of an exploration of the motivations and justifications of home experimenters. What follows considers the question of problem-solving – revealing how everyday, domestic issues prompted individuals to experiment with materials and techniques in the hope of sharing productive innovation with wider society. These examples incorporate those who sought knowledge about the natural world for its own sake and those who were mainly concerned with the potential for commercial gain. Many were motivated by a combination of the two. The examples are drawn from a diverse social pool, including barely literate workers alongside the professions and landed classes. This discussion connects with debates about the terms used to describe knowledge developed in different contexts and the hierarchies or dichotomies these terms can evoke, which will be discussed further in this book's Chapter 7.[6]

As discussed in Chapter 5, the archives of the Royal Society of Arts, formerly the Society for the Encouragement of Arts, Manufactures and Commerce, hold important insights into the investigative practices of a wide range of eighteenth-century men and women. The Society wished to encourage creative thinking that could be put to public use with a view to fostering beneficial social progress. Amongst the letters submitted to the Society for the Encouragement of Arts, Manufactures and Commerce and the American Philosophical Society describing efforts to innovate for the 'public good', there is weighty evidence that the home was the key site for this activity. Sometimes the ingenious new observation, adaptation or product was even prompted by a domestic problem. This is mirrored in other kinds of archival records; for example, in the household papers of the O'Hara family of Annaghmore, Co. Sligo, note was made of a technique of using powders, including 'Chrystals of Sulphate of Soda', placed in a 'Tin bottle holder' with cold water to chill a bottle of wine in twenty to thirty minutes.[7] In an era before the technology of refrigeration and when making, keeping and transporting ice was logistically challenging, no doubt this technique provided a welcome shortcut. Even when an innovation was not directly motivated by a domestic problem, the experimental practice of these letter-writers clearly displayed the knowledge and skills honed by work in the home, garden or field.

When Lewis Nicola wrote to the American Society in 1769, he described 'An easy Method of preserving Subjects in Spirits'.[8] This was an explicitly scientific project, as Nicola designed to maintain natural history specimens. He sought to guide his readers in strategies that avoided the disappointment of spirits evaporating from the container and leaving specimens vulnerable to decomposition.

The advice was informed by methods formerly published by the famous naturalist, R. A. F. de Réaumur; however, in Nicola's version, the use of common domestic equipment was described and the similarities between recipe book instruction on food preservation and natural historical preservation become clear.[9] For example, after acquiring some specialised stoppered glass containers for the specimens, Nicola advises that the glass vessel be 'secured by a piece of bladder or leather tied around it and the neck of the bottle'.[10] This strategy recalls the technique described in Mrs Baker's 1810 cookery book for the purpose of pickling walnuts, in which the 'pot' is double-sealed 'first with a Bladder, and outside that with Leather, that no air may get to them'.[11] For specimen preservation, an airtight seal was especially important and Nicola shares an innovation of his own making involving the use of 'some thin putty, the consistence of a soft ointment' to help form the seal.[12] Whilst mercury was also proposed as an appropriate chemical to enclose in the seal to further reduce spirit leakage, Nicola followed Réaumur's suggestion that 'nut oil, thickened to the consistence of honey' could act as a substitute 'by a long exposure to the air which will give it weight sufficient to sink in a weak spirit'.[13] For those naturalists who could not acquire a vessel with a glass stopper, two layers of bladder would suffice – secured tightly around the top. However, to get the seal to properly set, it was advised that the glass bottle should be turned upside down, and in recognition that 'many bottles will not stand on their mouths', 'wooden cups, turned with a broad bottom and a hollow' could be used to balance the upturned bottle within.[14] Nicola further suggested that the wooden cup could be filled with 'melted tallow, or tallow mixed with wax, until all the bladder or leather cover is buried in it'.[15] Tallow was rendered animal fat, commonly used in the home for making both soap and candles. Thus, from typical preserving seals made from leather and bladder to wooden cups and melted tallow, the everyday techniques, materials and objects of domestic life appear in this description of specimen preservation.

Nicola stated that his suggested adaptations of Réaumur's advice were intended to reduce the cost and ensure the 'easiness' of procuring the materials. Besides the glass vessel or bottle, these were all common domestic materials and things. Nicola's instructions also assumed that his reader would have these domestic articles to hand.

Moreover, this example strongly echoes the instructions provided by recipe books and the materials and objects listed in account books and inventories explored in Chapters 1 and 2.

Many other submissions to these societies not only used the material culture of domestic life but were also inspired by everyday household challenges. One such example was A. Curteen of Haverhill in Suffolk who wrote to the London-based Society in June 1756. Keen to establish his knowledgeability on the chosen subject, Curteen stressed that he had made 'manifold and repeated experiments' over the course of 'fourteen or fifteen years together' but understood that there were some obstructions to 'this great discovery … becomeing an universall good'.[16] The topic was preserving the flesh of animals, with a view to the product provisioning sailors during long voyages at sea. Following a critique of the common practice of salting the 'flesh of sheep', Curteen proposed an alternative method: drying meats 'under a covered roof but laid open on every side to the wind', which he felt both reduced the likelihood of flies getting to the meat and also reduced the 'smell and taste of putrefaction' present in some salted products.[17]

In a similar vein, 'A. B.' wrote in 1761 about his 'Observations on the process of manufacturing oils' with a procedure that would improve the quality of 'any kind of fish or seal oil, that pitrifid & stinking' and 'the drain oil called vitious oil'.[18] In the case of the former:

> When the oil is taken off from the dregs & brine: the dregs which swim on the brine should be taken off it also & put into another vessel of a deep form: & on standing, particularly if fresh water be added & stirred with them, nearly the whole remaining part of the oil will separate from the foulness: or to save this trouble the dregs when taken off may be put to any future quantity of oil that is to be edulcorated by this method. Which will answer the same end.[19]

But for 'vitious' oil that was even 'more putrid & foul', this process would remove the bad smell 'however stinking it may be' and adjust 'the brown colour … to a very light amber'.[20] The innovator referred to domestic practices when he commented that these oils might be used in lamps and referenced a kind of oil commonly known as 'Kitchen stuff'. However, the potential to use these methods in a manufacturing context was the object, thereby not only attending to the needs of the frugal housekeeper but also contributing to the prosperity of the nation.

Sadly, it seems that 'A. B.' did not receive the response he required from the Society; he would write a further three times about oil (including a letter on 10 May 1761 running to thirteen large sides). He also sent specimens, which he worried about: 'I am apprehensive that the specimen sent is not purified equally to what my process can effect, & that as it may probably if kept in a warm room be less perfected than when it was sent.' The letter-writer was anxious that he had not heard back from the Society and presumed 'nothing is hitherto decided with respect to' the proposal. In hopes of improving his standing with the Society, he 'sent another sample of crude & purified oil, which I fancy will be found more different from each other than the first'.[21] Thus, highly practical home experiments with useful household materials found themselves lodged with the Society, samples sent for inspection and lengthy persuasive explanations written out in pen and ink over many leaves of paper. The hope of recognition is tangible on the page, as is the apprehension of rejection.

Some submissions included detailed descriptions, diagrams and even models of a particular innovation. When Richard A. Clare wrote from Clarendon in Jamaica on 21 April 1799, addressing himself to the secretary of the Society, Samuel More, he enclosed a diagram to which he referred in the text of his letter. His communication was concerned with a new design of 'Still and Refrigeratory, calculated to save expence in the distillation and refrigeration of ardent spirits, at the same time that it renders these more pure than can be done by stills of the usual construction'.[22] He felt sure that distillation on the principles that he described 'may turn an advantage' and noted that he would 'esteem himself honoured' should the Society approve his design.[23]

The cross-referencing between the image and text pictured in Figure 6.1 allowed Clare to explain the design of his invention. He cautioned More that 'Every part of the apparatus must be made air tight.'[24] Unfortunately, a year and three months later, Clare was forced to write to confess that the still of his invention had 'By some accident … got leaky, admitting the air when the vacuum was made.' For the time being, he was waylaid: 'as I have little leisure from my business as a Surgeon &c, I have not as yet set myself to repair it; for you must know there are no workmen in this country to execute a thing of this kind'.[25] However, he fully

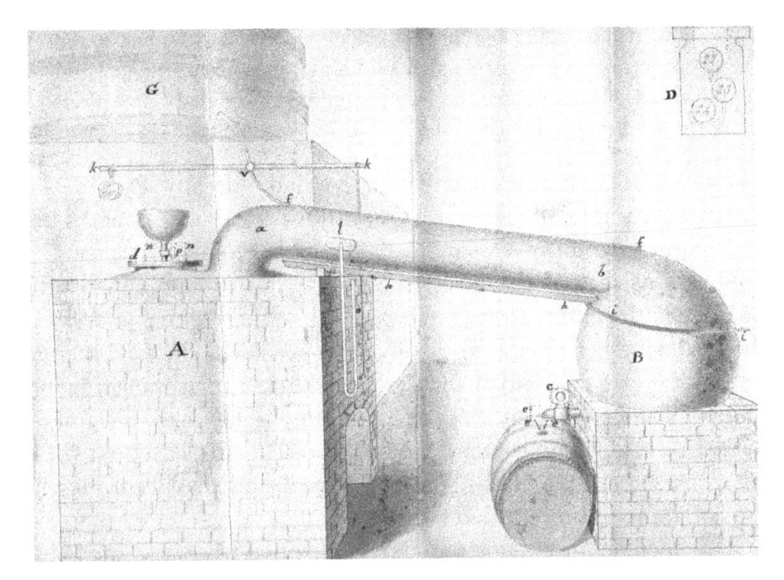

Figure 6.1 Richard A. Clare's 'Still and Refrigeratory'. Courtesy of Royal Society of Arts, London. All rights reserved and permission to use the figure must be obtained from the copyright holder.

intended, 'when I have time' to 'resume the inquiry, respecting the advantages that may arise from distilling in vacuo, and the result of my experiments shall be laid before the Society'.[26] Such correspondences with the Society could, in some cases, span years with willing experimenters sending updates on their observations and new adaptations that might be of interest to the arbiters of commercial and artistic merit.

Whilst many of the submissions to the Society came from the aristocratic, well-to-do or professional classes, this was certainly not the whole story. In fact, just as women correspondents dominated in the category of polite arts, manufacturers and merchants represented the majority of entrants in the fields of manufactures, mechanics and colonies and trade.[27] On 26 May 1791, a joiner – Alexander Thomson – living on the Nutts River Estate in Jamaica wrote to the Society about a mathematical instrument he had designed and made:

> I beg leave to Acquaint you that I have found out to Perfection A Mathematical Instrument (of my own making intirely of wood and

> made By myself being A joiner By Trade) which solves By Inspection all Quest[i]ons in Right Angled Obliq[u]e and Accute angles and likewise at pleasure solves Obliq[u]e and Accute Angled Quest[i]ons When Required to be reduced into two Right angles.[28]

Thomson assured the Society that he had 'Already Proved the Instrument and in all Cases and Quest[i]ons above mentioned finds it Accurate Both By Geometrical and Trigonometrical proof'.[29] Submissions such as this give a snapshot of the domestic ingenuity of eighteenth-century householders. Some of their letters addressed the challenges of the domestic environment itself, the vast majority reported on experiments undertaken in that environment using materials and equipment close at hand. All of them made a claim to authority in their understanding, whether that was based on their extensive experience, the application of well-honed skills or merely the commitment to improve.

The societies in London and Philadelphia were not the only organisations that captured the investigative activities of curious individuals from across the social spectrum. A list of patents issued in Shropshire over the course of the eighteenth century details products designed by a 'Gentleman', 'Yeoman', 'Clerk', 'Forgeman', 'Mathematician', 'Engraver', 'Ironmaster', 'Flax Dresser', 'Engineer', 'Clockmaker' and 'Coal Master'.[30] For the tradesmen in this cohort, innovations were often focused on improving the techniques of that trade; others used the specialised skills of their work to create something of use in an entirely different realm. For example, the yeoman Thomas Jackson of Wellington registered a 'Tincture for curing wounds, burns &c' in 1747.[31] Meanwhile, a mathematician and an engraver (John Duncombe and Joseph Pokle of Ludlow) collaborated to design a new method of measuring timber and a mechanical turning spit to replace a jack.[32] In these cases, the patents suggested a financial motivation alongside the more public-minded considerations often invoked in the societies' transactions. Nevertheless, they further corroborate the idea that a wide range of people engaged with innovation in technique in this period.

The examples discussed here speak to themes covered in previous chapters. Using a different range of primary sources, the importance of tacit knowledge, technique and the materials, objects, demands and affordances of domestic space still shine through. However, they also reveal the synergy between the home and a wide range of enquiry and innovation and hint at the personal investment individuals made in their own experiments, especially when they committed their knowledge to paper and asked an institution to acknowledge its worth. For some, the act of writing was a challenging one but the evidence of a wide range of literacies in the Society for the Encouragement of Arts, Manufactures and Commerce's archives is telling as to engagement from all classes. In the act of innovation, these home experimenters moved far beyond a narrow conception of know-how. The very fact that they were able to trial new methods showed their motivation to move beyond precepts learned through their trade or role and an urge to push the material limits of their surroundings. However, these testimonies largely focus on the invention or method itself and give fewer clues about the person behind the innovation. The home itself is most often implied rather than described. Next, the question of identity will be considered, including the ways intellectual authority was constructed by curious eighteenth-century experimenters.

Constructing female authority

Here the discussion moves to consider the way curious individuals might have viewed their own activities in a period of stark social and intellectual hierarchies. Whilst the Dublin Society was open to the idea of idle 'spinsters' putting their hands to work in the service of the nation, on the whole women were not expected to participate meaningfully in the more elevated projects of refining technique and producing knowledge. Moreover, as Ludmilla Jordanova has argued, 'as fields with a privileged relationship to nature', the natural sciences 'play a major role in explaining and disseminating gender as a naturalized category'.[33] For eighteenth-century women who had an interest in these subjects, the terms of engagement and the perceptions of their activity were necessarily filtered in specific and often prejudicial ways. Nevertheless, women did involve themselves in the natural sciences and navigated their own paths through the prescriptions of science and society, often co-opting or adapting gendered discourse to serve their own purposes.[34]

As discussed in Chapter 5, Ann Williams and Henrietta Rhodes evidenced their expertise by providing detailed reports of their observations and experiments with silkworms – bringing to bear the 'evidentiary weight of observation' in their submissions to the Society for the Encouragement of Arts, Manufactures and Commerce.[35] So did Elizabeth 'Wyndham', the mistress of an Earl. Whilst women remained excluded from roles within the learned societies and universities of eighteenth-century Britain and Ireland, the century did offer other inroads to scientific enquiry and writing. Building on activity undertaken by largely aristocratic women of the 1600s in the fields of experimental science, medicine and technical writing,[36] in the 1700s a more diverse range of women were engaging with science in public fora, whether that was through periodicals or poetry.[37] Prevailing pessimism about women's abilities to participate in intellectual life did not deter many women from entering this arena, either as a private domestic practice or as a documented – even published – scholarly pursuit.

When writing to the Society about their silkworm colonies, both Ann Williams and Henrietta Rhodes took care to present their findings as authoritative using a range of justifications for the conduct and conclusions of their work. Williams was particularly keen to offer a transparent account of her decision-making process: 'As to Cocoons, I have none, for after my first essay of reeling off about a dozen, I observed the silk, the nearer it came to the cocoons, grew finer, stronger, and better coloured. It immediately occurred why might not the whole cocoon be reeled off.'[38] To this end, she tried 'the experiment in water, so hot I could scarce keep my hand in', and it lived up to her hopes: 'The strong glutinous matter, which forms the contexture of the cocoon, immediately gave way, and I reeled off every single thread.'[39] However, the women positioned their activities very differently from one another and their language reflected distinct constructions of female authority in relation to knowledge and skill.

Williams situated her enquiry as partly a natural historical one, commenting that she believed 'that half the benefit arising from this minute part of the grand Creator's works are not yet unravelled'.[40] In a long letter dated 19 October 1777, Williams described the worms as they were about to produce silk, noting that the first indication was 'a transparency all over them, with a visible circulation of the blood, or glutinous matter'.[41] Williams 'humbly' inferred that this action 'forms the silk, and assists the spinning', adding that the substance 'is visibly seen circulating down the middle of the back'.[42] Next, she observed that

> they erect themselves on their bellies, with their heads in form of a sphinx, sometimes seeming to play, biting their sides and silken tail, then lying dormant: But the most certain criterion is, when they eat from side to side of the large fibres in a circular form, nibbling the leaves to atoms, and wasting them. At this period, they become of a fleshy colour, their backs appear very luminous, especially by candle light.[43]

This passage of close observation is concluded with a mention of the silkworms moving 'in a circular manner from side to side of the box' and the more practical assertion that 'at this moment they are to be put in papers or all the labour will prove abortive'.[44]

By delivering such detailed anatomical descriptions, Williams was contributing to a broader culture of women both collecting and documenting flora and fauna in this period. Famous examples included the exceptional naturalist and illustrator, Maria Sibylla Merian (1647–1717), and aristocratic collectors such as Margaret Cavendish Bentinck, Duchess of Portland, discussed in Chapter 3, but these were modes that could be adopted by lower-profile and lower-status individuals.[45] Whilst Williams did not lay claim to the title of naturalist herself, she saw her work as making a solid contribution to this realm of scholarly activity. She upheld the importance of her own domestic observations: 'but this I know, which is well worth the while of naturalists to investigate, that the female Aurelia is full of eggs before she changes her state to that of a Chrysalis'.[46]

By contrast, Henrietta Rhodes emphasised the wider significance of her work in terms of the economic potential of large-scale manufacturing:

> I am decidedly of opinion, that this great article of commerce, which use and luxury have rendered so essential to our comforts and conveniencies, and for which such immense sums are annually sent into other nations, may be cultivated at home with the greatest ease, and with the utmost certainty of success.[47]

She argued that 'from the recital I have given', it is clear that thirty thousand silkworms would be required to produce five pounds of silk. She further reasoned that as twelve large mulberry bushes 'were scarcely adequate to the support of ten thousand' in her possession, 'any means to stimulate the spirit of making Mulberry plantations' would be critical to success.[48] She herself had managed ten thousand 'with ease and success', but she advised that if others were to follow in her steps and on a larger scale, 'the expence of erecting a place for them would be very trifling' but they would need two people to attend the enterprise.[49] These concerns were echoed in the Dublin Society's records, as three women explicitly referred to their need to take on extra help with their silkworm colonies, whether that was through training other women like themselves or by employing children from 'the Public Schools' as apprentices.[50]

Rhodes also emphasised the ways in which her investigation responded very closely to the Society's objective of developing practical knowledge. Whilst her letters occasionally evoked regimes of domestic care, her discourse emphasised problem-solving over nurture.[51] Rhodes's original motivation had been to produce 'the quantity of Silk necessary for a dress' and, whilst the subsequent years of experimentation allowed her to draw far broader conclusions, this interest in manufacture sat at the heart of her project.[52]

Rhodes used comparisons with her predecessor, Williams, to prove the pre-eminence of her practice. In defending the lower yields of her silkworms, she accused Williams of taking 'waste or carding silk into the account', a habit that Rhodes regarded as 'incompatible with my ideas of truth and candour' and out of line with the intentions of the Society when they offered their premiums.[53] Rhodes's criticisms of her predecessor's reports offered an implicit contrast with her own work. This was important because Rhodes was asking the fellows of the Society to trust the integrity of her practices and observations thereof. She also referred to measuring the silk from a cone 'with the most critical exactness'.[54] Accuracy of practice played an important role in both women's efforts to present their accounts as authoritative. Many more criticisms of Williams's work exist in Rhodes's letters, as she weighed up the range of evidence she had read in the Society's *Transactions* against her own experience. By demonstrating a marked improvement on a prize-worthy submission by another female applicant, Rhodes sought to concretise her own achievements in the eyes of the Society.

The language used by the women is also telling. Like Moses Bartram, they anthropomorphised the silkworms. For example, both referred to their silkworm colonies as 'families' and Rhodes described her silkworms as 'industrious little animals who depend on me solely'.[55] Overall, however, it is Williams's letters that most commonly deploy anthropomorphic references. For example, she stressed the importance of treating the silkworms with kindness and drew connections between her care for them and their productivity:

> I do not approve of the method used ... of striking them with a feather off the leaves to which they strongly adhere, as every time that practice is used, they not only lose a quantity of silk, but are visibly in pain, which may be seen by their various contortions; by these means, and keeping them dirty, they do not rear one tenth part of what they hatch, nor bring them to any size.[56]

Whilst she had clearly learned from published guides on the cultivation of silkworms, in the final analysis, she trusted her own direct experience. She sometimes referenced her own embodied knowledge to make her point, for example when reporting that she 'only used milk warm water, in the first process'.[57] In this way, Williams built the case for her own success through language that evoked female regimes of care.

Williams's approach corresponded with trends in science writing of this era, not least – as Londa Schiebinger has argued – the use of 'explicitly anthropomorphic thinking' in relation to botany 'ascribing to plants human form, function, and even emotion'.[58] More than once, Williams referred to her silkworms as 'my little family'; she inferred from their behaviours that the silkworms were 'innocent', 'satisfied' or in 'pain'; she noted when she thought they seemed to 'play' and when they reacted with 'horror'. She witnessed them seem 'satisfied' with their food and 'nestle into the pipes and repose themselves'.[59] Her language finds a reflection in the writings of the poet Anna Barbauld. In the 1770s, Barbauld (1743–1825, and at this time Aikin) commented on her friend and natural philosopher Joseph Priestly's experiments with mice, writing 'The Mouse's Petition'.[60] This poem was written in the voice of the mouse and made a plea to Priestley to release him. As Mary Ellen Bellanca argues, the poem 'does not simply inscribe a showdown between scientific patriarchy and feminine sensibility' as Barbauld fully supported scientific experiment and advance, but it does reveal her interest in promoting compassion towards animals – a cause that attracted support in her own era as well as later.[61] Barbauld's wider work took up a position that, all at once, promoted scientific knowledge for both men and women, aimed to 'reinforce cultural boundaries between the sexes' intellectual territories' and cautioned against the 'excessive ambition of male scientists'.[62] Whilst Williams's narrative assumes the legitimacy of her activities, the combination of authoritative observation and anthropomorphic allusion is striking. By engaging with the pain or comfort of her silkworms, emotion and empathy formed a part of her scientific narrative.

First-hand accounts of silkworm rearing took centre stage in both these women's letters, but occasional comments reveal that each of them discussed their venture with others and had read published accounts of the process. Local people, friends and experts were all referenced for the purpose of corroborating the efficacy of their approach. As discussed in Chapter 5, Rhodes was particularly well networked amongst the wealthy who could support her intellectual endeavours.[63] She had also read widely on this topic, mentioning on 24 August 1785 a 'Treatise' she had digested concerning plans to establish a silk manufactory in Georgia, the 'ingenious hint' of the 'Honourable Daines Barrington' on collecting leaves, which she had probably found in the Society's own *Transactions*, and her disagreement with the French Jesuit Jean-Baptiste Du Halde's judgement that noise was 'prejudicial to the Silk-worm'.[64]

Williams occasionally mentioned other people in her acquaintance, noting on 14 October 1777, 'Every person here, those who have kept them, as well as others, will have it that I have performed a miracle.'[65] Whilst it seems that Williams lived alone after the death of her father, she had clearly discussed her project with neighbours or friends, some of whom had tried silkworm rearing for themselves. Five days later, she reported, 'A Gentleman has been at my office, who lived three years in Italy [where silkworms were commercially reared], he declared though he had seen many thousands spin there, he never saw finer Worms than mine, and expressed his astonishment at their spinning at this season [October].'[66] This comment points to the community of interest that had cohered around the subject of silkworms and the access Williams had to information about silkworm rearing in other parts of the world. It also suggests that she had developed enough of a reputation for her work in this field that visitors to the area might drop by to examine her colony and approach.

Just as Williams's letters to the Society had no doubt prompted others to investigate, Henrietta Rhodes's 'elegant letters' were referred to in a subsequent submission by a Mr Swaine of Puckleworth near Bristol. He mentioned that the 'letters of that ingenious young lady' had induced him to write to the Society not 'in the light of rivalship; but merely to corroborate the testimony there adduced'.[67] It is worth also noting that Rhodes's writings enjoyed a considerable afterlife, being re-printed into the nineteenth century.[68] In this way, Williams and Rhodes participated not only in local communities of experimenters but also in a growing and diverse network, not of friends and relations, but of investigators reporting their findings to an institution, with the hopes both of a prize and the honour of contributing to this public project. Their interventions influenced others for decades.

Mastery of print culture

The eighteenth century experienced an exponential growth in print culture, cheaper print products proving particularly successful, and a significant rise in literacy that reached women and working people.[69] Developments such as the circulating library also expanded access to knowledge in the form of print.[70] Whilst the aristocratic, private library was still a major advantage to enquiry, the lack of it was no complete barrier. As discussed in Chapter 4, working men with little spare time or resources to aid enquiry made good use of the periodical press to corroborate their own findings and compare those of others.

Many scientific subjects graced the pages of cheap print in this period, the *Gentleman's Magazine* boasting in December 1808 that it was the premier publication for 'literary and scientific men to obtain information, on any particular subject in which they may be interested'.[71] Indeed, the *Gentleman's Magazine* routinely advertised the premiums of the Society for the Encouragement of Arts,

Manufactures and Commerce, presumably to ensure a broader audience than the Society's own *Transactions*.[72] The *Magazine* featured running topics, with submissions from multiple interested individuals, often responding to each other's ideas over a period of months, and sometimes years.

For example, throughout 1740, submissions about optical experiments involving mirrors were printed. One 'G. S.' wrote that an 'Optical Phenomenon is not yet taken notice of by the Writers on that Subject', the author rationalising that 'if propos'd in your Mag' it might reach the notice of '*Literati*' who could explain it 'from their Principles'.[73] He or she had been surprised when 'Looking at the Moon (by accident) in a common plain Mirror or Looking-Glass', 'to see her multiply'd into four distinct Spectrums, at some distance from each other'. G. S. tried the same experiment with the sun with a similar effect and wanted to know 'How is this to be accounted for in a plain, polish'd Mirror, where other Objects appear only single, as daily Practice confirms?'[74] In June of the same year, G. S. returned to the same subject, referring to a Mr Martin and his 'many curious Experiments ... relating to the optical Phenomenon'.[75] This Mr Martin was most likely Benjamin Martin (1704–82) who was a schoolmaster, optician and instrument-maker who gave public lectures and published on his expertise.[76] On the question of the 'Plurality of Spectrums' visible 'in the common Mirror', G. S. bemoaned that 'all the experimental Variety' of Martin was not enough to 'render a decisional Solution' to this phenomenon.[77] G. S. elaborated a list of eighteen observations about these spectrums and a diagram, suggesting that this amounted to an explanation of the phenomenon and invited Martin or other 'Proficients in that Science' to concur or elaborate an alternative.[78]

As it happened, on 1 August 1740, a P. Wood of Cheshire responded to this call with his or her own observations.[79] The submission was aimed at the 'optical Readers' of the *Gentleman's Magazine*.[80] As with the former contributor, a list of numbered observations and a diagram followed and the piece concluded with a statement agreeing with 'Mr Martin' 'that several Difficulties do attend the Nature of this question, which must be accounted for otherwise'; nonetheless P. Wood had hoped to contribute to the discussion and elaboration of this phenomenon, using the *Magazine* as a means of conversing with other people who were curious about optics, not least the famous instrument-maker.[81]

As discussed, astronomy was a subject that appeared regularly in the pages of the *Gentleman's Magazine*, alongside a wide range of other periodicals of this period. Curious individuals wrote into the *Magazine* from a range of standpoints. One contributor described himself in March 1742 as 'entirely a Novice in all Parts of mathematical Learning' and enquired about the path of the moon. The following month, a direct response from a more experienced astronomer, calling themselves 'Philalethes' from County Durham, was printed.[82] The diverse participation in this pursuit is exemplified by the fact that the observations of internationally known astronomers were listed beside the anonymous initials of interested individuals. For example, the solar eclipse of 4 August 1739 was reported in the September edition of the *Magazine*, which gave the observations of the first astronomer to the Empress of Russia, I. N. De L'Isle's sighting, alongside 'I. B.' of Stoke Newington and 'J. T.' of Newcastle-on-Tyne.[83]

Those who wrote to the *Gentleman's Magazine* were not all working with the same tools at hand. For example, 'J. B.' made 'Observations of the Occult action of Jupiter by the Moon' from Fleet Street, London on 27 October 1740 with 'an excellent reflecting Telescope which magnify'd 120 Times'.[84] By contrast, on 26 February 1742, Thomas Wright of St James's in London thanked the *Magazine* for an account of a comet, which Wright had sighted himself at three o'clock in the morning. However, 'for Want of proper Instruments to observe it', he used a length of thread to determine the position of the comet in relation to longitude and latitude, a technique he helpfully shared with other readers. This inexpensive method, he assured them, would allow 'the Place of the Comet' to be 'very easily found'.[85] On the same page of the *Magazine*, a G. Smith of Boothby, near Carlisle, similarly professed to spot the comet despite having 'no Instrument to make proper Observations'.[86] This form of positional astronomy was clearly open to very many, without the need for expensive instruments, and formed a mainstay of astronomical discussion between the Dublin apprentices, Jackson and Chandlee, discussed in detail in Chapter 4.

As Jackson and Chandlee's letters reveal, the periodical press provided important information and Jackson habitually annotated copies with his own observations. Via his father's print shop, he had a broad range of print culture within his reach. This was a resource he often shared with his friend, although one he maintained an exacting control over, writing on 7 May 1769, 'Thine receiv'd, which informs one that thou art very covetous of reading Magazines – but there is one material thing that thou hast not yet considered viz. that a Bookseller above all people hate to lend Books, except he makes a trade of it.'[87] Of course, as a trainee bookseller, Jackson was satirising his own attitude to lending his reading material, but the letters regularly refer either to the loan of a particular item or the reasons why it could not be borrowed, often in a tongue-in-cheek style. For example, Jackson offered Chandlee a 'view of the Cambridge Magazine', but elaborated,

> there are different sorts of views some transient, and others of a longer continuance, I fear that while one person would be taking a long view of it in one street, another person in the place it came from, might want to peruse or review it; and here might fall on a great inconveniency.[88]

However, on occasion, the apprentice to a linen-draper Chandlee was able to lend Jackson a magazine. For example, Jackson asked his friend 'Wilt thy have thy old piece of a *Lady's Almanack*?', complaining that 'It helps encumber my desk; and I don't want it.'[89] Despite the off-hand mode of expression, this shows that Jackson and Chandlee also read publications explicitly marketed at women; however, such periodicals could easily have featured subjects of key interest as mathematical problems were a mainstay for publications such as *The Ladies' Diary*, also known as the *Woman's Almanack*.[90] Both men were clearly familiar with a wide range of periodicals and Jackson also sourced publications from beyond Ireland's shores, noting on 23 April 1769, 'The long time we have to wait sometimes and the Extraordinary loss of shoe leather are great discouragements to taking the English Magazines.'[91]

Whilst the apprentices' scientific interests drove some of their engagement with print culture, Jackson's training to become a printer and publisher was also influential and this topic sometimes interrupted the flow of astronomical exchange. In August 1769, he wrote a postscript to his letter commenting, 'This year hast been a very remarkable year, for the breaking of Printers, 3 have had their good sold by Auction &c.'[92] Each had their own financial vices; one was also 'a stage player' who 'went in debt to some of the Playhouse folks', another was too fond of 'taking in new expensive household furniture, than he was at paying for them' and the third was experienced in the trade but 'so unwise as to practise Gaming'.[93] Jackson was also critical of some common practices among booksellers, reporting 'piratical depredations ... committed in the Almanacks way'.[94] Noting one outlet's 'annually stolen sheet almanac which is no new piracy', Jackson bemoaned another 'which appears to be almost entirely copied from Watson', meanwhile 'B. Corcoran published a base and servile Imitation of Smith's sheet almanac under the title of the Royal Dublin Sheet Almanack'.[95] Jackson ended his letter with the comment 'Some folks it seems do not so much consider what is fair and just, and agreeable to the Royal Law, as what will bring Profit.'[96] In this way, Jackson – and increasingly Chandlee – were very well acquainted with the contents of the periodical press, both local publications and English imports, but also knowledgeable about the business and the poor habits of some printers in reproducing content wholesale from one almanac to the next.[97] This insider knowledge and ability to secure diverse publications gave these young men a real mastery when it came to assessing the value and reliability of information sourced from cheap print.

As discussed in Chapter 4, Robert Jackson did not always trust the astronomical content of almanacs. On 24 December 1769, he disagreed with 'Two of our Irish almanack writers' who were predicting a visible eclipse in 1770. As much as Jackson 'should like to see it as well as any of them' he had 'good and sufficient reason to believe' that this would not come to pass.[98] In a long letter, Jackson included a final section entitled 'Intelligence' in which he imparted his criticism of a new almanac printed in Cork by a 'John Scanlon of Cloyne'. Issues included Scanlon's 'shameful still-continued Blunder in the Tides', his neglecting to record 'the invisible Eclipse in 1st mo. [January]' and the accusation that he 'hath laid violent hands on the Transit of Venus, and instead of placing it on the 3rd of June partly visible (till the sun setting deprives us of it)', he had printed the date as 4 June 'and the greatest part of it visible, after the sun's rising; which I have reason to apprehend is quite false'.[99] After some further comparisons between printed astronomical details, Jackson concluded by admitting 'I am not yet ready to take a Critical review of the Eclipsio-graphers and transitographers for 1769', suggesting he might do so in five or six weeks' time.[100] The 'Eclisio-graphers' most likely referred to those who produced the *Nautical Almanac*, or 'eclipsiography', because longitude could be calculated by comparing the local time of a lunar eclipse at two distant places.[101] Astronomer Royal, Nevil Maskelyne, worked on a new *Nautical Almanac and Astronomical Ephemeris* for the year 1767 and published it in 1766.[102] This process of reading, comparing and – often – critiquing and disputing the published details of eclipses, transits and other astronomical phenomena was a mainstay in these letters and demonstrates the active and analytical quality of these young men's reading practices.

Jackson and Chandlee's letters illuminate a readership for the periodical press with both expertise in the published subject and the genre itself. This calls into question the notion that periodicals were merely channels for 'diffusing a taste for those useful sciences over the nation at large' as the *Irish Magazine for Neglected Biography* of 1810 described it.[103] This book argues that there is much more agency in the so-called 'lower orders' than that traditional characterisation of information flow would account for. As Jackson himself was engaged in the process of printing and selling print culture, he perhaps felt a greater authority to produce, consume and challenge its content but – more than anything else – his letters reveal that the information disseminated in these almanacs was seized upon by a selection of working people in the city who could form their own views on the validity of what they read. With this knowledge came a power over print and an authority in the critique of it. Like the individuals who contributed letters, observations and experiments to the *Gentleman's Magazine* discussed above, these working men were not passive consumers, but active participants in a diffuse intellectual culture, a culture of the curious in eighteenth-century urban life.

Conclusion

The domestic experimenters discussed here presented their personal experiences as authoritative in the framework of scientific observation and as useful to commercial innovation, but they articulated their authority in different ways. Cognition is an emotional act and emotional worlds not only motivate but also constitute knowledge-making.[104] The affective resonance of home and the common child-hood experience of learning at a carer's side create the conditions for first understanding the world. Whilst the individuals discussed here rarely foregrounded the emotional, in case it detracted from the rational, they did betray the importance of their scientific work to their sense of self, sometimes in heartfelt terms.

Ann Williams and Henrietta Rhodes were women of the middling sort and gentry who considered their own domestic experience of breeding silkworms and harvesting their crops as valid evidence on account of the 'truth and candour' of their testimony. Williams's testimony drew strongly on concepts of nurture and domesticity, extending the discourse of familial care to encompass this task and simultaneously developing a vocabulary for describing and interpreting her findings. Whilst Rhodes chose to prioritise the language of production over that of care, like Williams, she conducted her activities at home and transferred materials, equipment and technique from one domestic task to the next. Ultimately, the cultivation of these living creatures formed one part of the material and social life of the home, which included the care of family members and the careful stewardship of domestic resources.

Robert Jackson and Thomas Chandlee filled their correspondence with the details of their shared enquiry, in the process revealing their close engagement with a prevalent form of eighteenth-century print culture. Jackson's use of the playful pseudonym 'Philalithes Astronomus' indicates both identification with his chosen pursuit but also the humour of shared endeavour that runs through this correspondence. Using such pseudonyms was a mainstay for correspondents with the periodical press and, in doing so, Jackson underlined his place in that culture. As Chandlee's tutor, Jackson often assumed a didactic tone and wrote with assured confidence about a range of astronomical questions, methods and phenomena. However, it is in the encompassing command of periodical output that Jackson's authoritative persona really takes flight – marrying as it does his burgeoning professional identity with his subject specialism.

Active participation in print culture, both as critical readers and knowledgeable or curious contributors, formed a bedrock for independent observation and experiment. Whilst some correspondents with formal societies clearly feared rejection of their proposal, many others had a firm confidence in the value of their innovation, underpinned by personal expertise honed through experience. They often articulated this self-confidence. In many submissions to societies, the particulars of the home and the correspondent's own sense of self have to be read between the lines. However, examples such as Williams and Rhodes and Jackson and Chandlee give an authoritative voice to the variegated expertise that could be sharpened at home. Their confidence was bolstered by membership in multiple communities of enquiry, in ink, print and person. Taken together, their testimonies indicate widespread participation in scientific knowledge-making by a diverse population of individuals. The next chapter argues that a culture of curiosity prevailed in the eighteenth century and, in doing so, re-examines the ways historians characterise intellectual life in this era of 'Enlightenment'.

Notes

1 Shapin, *Social history of truth*.
2 For example, Philippa Hellawell, '"The best and most practical philosophers": Seamen and the authority of experience in early modern science', *History of Science*, 58:1 (2019), pp. 1–23.
3 *Gentleman's Magazine*, vol. 64, pt. 2 (Jul. 1794), p. 638.
4 *Ibid.*
5 *Ibid.*, p. 640.
6 See, for example, these titular formulations: Leong, *Recipes*; Susan Whyman, *The useful knowledge of William Hutton: Culture and industry in eighteenth-century Birmingham* (Oxford: Oxford University Press, 2018).
7 NLI, O'Hara of Annaghmore Papers, MS 36,375/3.
8 Lewis Nicola, 'An easy method of preserving subjects in spirits', *Transactions of the American Philosophical Society*, vol. 1 (1769–71), pp. 244–6. 'Spirits' generally referred to some form of alcohol including rum, gin or brandy; see Robert McCracken Peck, 'Preserving nature for study and display' in Sue Ann Prince (ed.), *Stuffing birds, pressing plants, shaping knowledge: Natural history in North America, 1730–1860* (Philadelphia, PA: American Philosophical Society, 2003), p. 13 (pp. 11–25).

9 According to Lewis Nicola, Réaumur had published his methods in the 1746 edition of the *Memoirs of the Royal Academy of Sciences*.
10 *Transactions of the American Philosophical Society*, vol. 1 (1 Jan. 1769–1 Jan. 1771), p. 244.
11 NLI, 'Mrs A. W. Baker's Cookery Book, 1810', MS 34,952, vol. 1, fo. 10; see Chapter 2.
12 *Transactions of the American Philosophical Society*, vol. 1 (1 Jan. 1769–1 Jan. 1771), p. 245.
13 *Ibid.*, p. 244.
14 *Ibid.*, p. 245.
15 *Ibid.*
16 RSA, PR/GE/110/5/18: 18 Jun. 1756.
17 *Ibid.*
18 RSA, PR/GE/110/11/2: 1761.
19 *Ibid.*
20 *Ibid.*
21 RSA, PR/GE/110/11/25: 13 Apr. 1761.
22 RSA, PR/MC/105/10/469: 21 Apr. 1799.
23 *Ibid.*
24 RSA, PR/MC/105/10/470: 20 Jul. 1800.
25 *Ibid.*
26 *Ibid.*
27 For this information I am grateful to Anton Howes whose research on the RSA revealed these class biases according to category; agriculture courted the highest levels of interest from gentry and even nobility, but this was not the picture across the piece.
28 RSA, PR/MC/101/10/468: 26 May 1791.
29 *Ibid.*
30 SA, 'A list of patents granted under the old law 1617 to 1852', C20/2629/1.
31 *Ibid.*
32 *Ibid.*
33 Jordanova, 'Gender', p. 482, this piece further argues that the potential of gender as an analytical tool for the history of science can only be realised if it is treated comparatively and contextually.
34 See, for example, Ann B. Schteir, *Cultivating women, cultivating science: Flora's daughters and botany in England, 1760 to 1860* (London: Johns Hopkins Press Ltd., 1996); Sam George, *Botany, sexuality and women's writing, 1760–1830: From modest shoot to forward plant* (Manchester: Manchester University Press, 2007).
35 Daston and Lunbeck, *Histories of scientific observation*, p. 115; for the relationship between observation and the self, see Lorraine Daston

and Peter Galison, *Objectivity* (New York: Zone Books, 2007), pp. 191–251.
36 See Hunter and Hutton, *Women, science and medicine*.
37 Costa, 'The "Ladies' diary"; Donna Landry, 'Green languages? Women poets as naturalists in 1653 and 1807', *Huntington Library Quarterly*, 63:4 (2000), pp. 467–89.
38 RSA, *Transactions*, vol. 2 (1784), p. 163.
39 *Ibid.*, p. 164.
40 *Ibid.*, pp. 167–8.
41 *Ibid.*, pp. 159–60.
42 *Ibid.*, p. 160.
43 *Ibid.*
44 *Ibid.*
45 See Boris Friedewald, *A butterfly journey: Maria Sibylla Merian artist and scientist* (Munich: Prestel, 2015) and Beth Fowkes Tobin, *The Duchess's shells: Natural history collecting in the age of Cook's voyages* (London: Yale University Press, 2014).
46 RSA, *Transactions*, vol. 2 (1784), pp. 164–5.
47 RSA, *Transactions*, vol. 4 (1786), p. 158.
48 *Ibid.*, p. 155.
49 *Ibid.*
50 Elizabeth Cortez trained up a Mrs Anna Bell and Martha Charlotte Menzies was keen to secure apprentices, see *Proceedings of the Dublin Society*, vol. XII (Nov. 1775–Jun. 1776), p. 3 and vol. XXXIX (4 Nov. 1802–11 Aug. 1803), p. 2.
51 RSA, *Transactions*, vol. 4 (1786), p. 150.
52 RSA, *Transactions*, vol. 5 (1787), p. 146.
53 RSA, *Transactions*, vol. 4 (1786), p. 162.
54 *Ibid.*, p. 163.
55 *Ibid.*, p. 150.
56 RSA, *Transactions*, vol. 2 (1784), p. 159.
57 *Ibid.*, p. 164.
58 Londa Schiebinger, 'Gender and natural history' in Jardine, Secord and Spary, *Cultures of natural history*, p. 170 (pp. 163–77); this style contributed to the interest in sexual difference that was developing in the eighteenth century.
59 RSA, *Transactions*, vol. 2 (1784), p. 157.
60 Mary E. Bellanca, 'Science, animal sympathy, and Anna Barbauld's "The mouse's petition"', *Eighteenth-Century Studies*, 37:1 (2003), pp. 47–67.
61 *Ibid.*, p. 49. For example, Mary Wollstonecraft appreciated Barbauld's line of argument.

62 Bellanca, 'Science, animal sympathy', p. 49.
63 Rhodes's *Poems and miscellaneous essays* (Brentford, 1814) was published by subscription revealing a large number of wealthy supporters.
64 RSA, *Transactions*, vol. 4 (1786), pp. 162–3, 164, 167.
65 RSA, *Transactions*, vol. 2 (1784), p. 155.
66 *Ibid.*, p. 161.
67 RSA, *Transactions*, vol. 5 (1787), pp. 150–1.
68 Paskins, 'Sentimental industry', p. 115.
69 See Bob Harris, 'Print culture' in H. T. Dickenson (ed.), *A companion to eighteenth-century Britain* (Oxford: Blackwell Publishing, 2002), pp. 283–93; Toby Barnard, *Brought to book: Print in Ireland, 1680–1784* (Dublin: Four Courts Press, 2017). Literacy studies have moved away from signatures as the acid test for basic written literacy and towards more multi-layered assessments; see Eleanor Hubbard, 'Reading, writing, and initialing: Female literacy in early modern London', *Journal of British Studies*, 54:3 (2015), pp. 553–77; Steven Cowan, 'The growth of public literacy in eighteenth-century England' (PhD thesis, Institute of Education, University of London, 2012); on epistolary literacy see Susan Whyman, *The pen and the people: English letter writers, 1660–1800* (Oxford: Oxford University Press, 2009), pp. 9–11; on reading practices see James Raven, Helen Small and Naomi Tadmor (eds), *The practice and representation of reading in England* (Cambridge: Cambridge University Press, 2007); Mark R. M. *Towsey, Reading history in Britain and America, c.1750–c.1840* (Cambridge: Cambridge University Press, 2019).
70 Eleanor Lochrie, 'A study of lending libraries in eighteenth-century Britain' (MSc dissertation, University of Strathclyde, 2015).
71 *Gentleman's Magazine*, vol. 78, pt. 2 (Dec. 1808), p. 1056; for more on this publication's history and content see Emily Lorraine de Montluzin, *Daily life in Georgian England as reported in the Gentleman's Magazine* (Lampeter: The Edwin Mellen Press, 2002); and Williamson, *British masculinity*.
72 See, for example, *Gentleman's Magazine*, vol. 68 (May 1790), pp. 453–60.
73 *Gentleman's Magazine*, vol. 10 (Mar. 1740), p. 130.
74 *Ibid.*
75 *Ibid.*, p. 298.
76 David A. Goss, 'Benjamin Martin (1704–1782) and his writings on the eye and eyeglasses', *Hindsight*, 41:2 (2010), pp. 41–8.
77 *Gentleman's Magazine*, vol. 10 (Jun. 1740), p. 298.
78 *Ibid.*, pp. 298–9.
79 *Gentleman's Magazine*, vol. 10 (Sep. 1740), pp. 451–2.

80 *Ibid.*, p. 451.
81 *Ibid.*, p. 452.
82 *Gentleman's Magazine*, vol. 12 (Apr. 1742), p. 210; *Gentleman's Magazine*, vol. 12 (May 1742), pp. 264–5; 'Philalethes' was a popular pseudonym and was an Ancient Greek name meaning lover of truth; see Chapter 4 on Robert Jackson's use of a similar pen-name.
83 *Gentleman's Magazine*, vol. 10 (Feb. 1740), p. 80.
84 *Gentleman's Magazine*, vol. 10 (Oct. 1740), p. 517.
85 *Gentleman's Magazine*, vol. 12 (Feb. 1742), pp. 106–7.
86 *Ibid.*, p. 106.
87 FHLD, Fennell, MSS Box 27, folder 1, letter 22: Jackson to Chandlee, 7 May 1769.
88 *Ibid.*
89 FHLD, Fennell, MSS Box 27, folder 3, letter 123: Jackson to Chandlee, n.d.
90 Costa, 'The "Ladies' diary"'; it is possible *The Ladies' Diary* was, in fact, the publication Jackson referred to in his letter.
91 FHLD, Fennell, MSS Box 27, folder 1, letter 21: Jackson to Chandlee, 23 Apr. 1769. Jackson noted acquiring an American almanac with a view to comparing its contents with more local publications; see FHLD, Fennell, MSS Box 27, folder 3, letter 123: Jackson to Chandlee, n.d.
92 FHLD, Fennell, MSS Box 27, folder 1, letter 26: Jackson to Chandlee, 12 Aug.; postscript added 20 Aug. 1769.
93 *Ibid.*
94 FHLD, Fennell, MSS Box 27, folder 1, letter 28: Jackson to Chandlee, 7 Oct. 1769.
95 *Ibid.*
96 *Ibid.*
97 It is worth noting that Dublin booksellers often imported a large proportion of their stocks from England in the earlier part of the period, but increasingly re-printed English texts as they were unhampered by the British Copyright Act of 1710. British periodicals were also commonly imported to Ireland; see Benson, 'Irish trade', pp. 368–9, 372.
98 FHLD, Fennell, MSS Box 27, folder 1, letter 34: Jackson to Chandlee, 24 Dec. 1769.
99 FHLD, Fennell, MSS Box 27, folder 3, letter 94: Jackson to Chandlee, n.d.
100 *Ibid.*
101 See an example of individuals calculating the longitude of Kingston, Jamaica, from measurements of eclipses and reporting them to the Royal Society: James Caitlin and James Short, 'An account of the

eclipse of the moon, June 8 1750', *Philosophical Transactions*, 46:496 (1749–50), pp. 523–5.

102 Maskelyne's sister, Lady Margaret Clive, is discussed in detail in Chapter 3.

103 Anon., 'An historical account of Irish almanacks', *Irish Magazine, or Monthly Asylum for Neglected Biography* (Jan. 1810), p. 477.

104 Deborah R. Coen, 'The common world: Histories of science and intimacy', *Modern Intellectual History*, 11:2 (2014), pp. 437–8 (pp. 417–38).

7

Re-examining the culture of enquiry

In 1773, postmistress and silkworm experimenter Ann Williams published a volume of ninety-five poems, songs and riddles under the title *Original poems and imitations*. Oddly, not one piece dwelt on the topic of silkworms although the book itself was dedicated to the 'Right Honourable H. F. Thynne, His Majesty's Post-Master General', underlining the author's identification with her occupation.[1] The contents cover an eclectic range of subjects, including friendship, death, the structure of the human body, earthquakes and astronomy. They celebrate, among others, Shakespeare, Virgil and General James Wolfe.[2] Many poems take the form of an address to a gentleman on questions of womanhood, such as, 'Familiar epistle to a gentleman who wrote a bitter satyr against women' or 'Impromptu, to a gentleman who railed against the ladies, particularly the married ones'.[3] Williams's cultural and scientific interests are on display here, alongside her thoughts about female intellect. Not only do some poems explain to a gentleman the value of female learning, but others consider the act of thought itself.[4] Several poems respond to famous women writers who Williams clearly admired – for example, 'On reading some lines in praise of Mrs. Macauly' and 'Impromptu, on reading Mrs. Rowe's poems', the latter occupying the last pages of the book.[5]

Williams describes the poet, Elizabeth Singer Rowe, as an inspiration, noting the deadening effect of some formal education upon creativity:

> Behold a pattern here for womankind!
> By nature deck'd with an exalted mind;
> Who wrote from her, not by pedantic rules
> Of musty morals, taught in rigid schools.[6]

These sentiments speak to Williams's own status as an autodidact. Whilst it is difficult to determine her experiences of formal education in childhood, the extant records show that she advanced her own understanding of a range of fields informally and in the context of her job. The poem implies that free of the 'pedantic rules' of school, the poet and intellectual could reach her full potential.

Next, Williams dwelt on a feature of her own publication – the imitation of other great writers' work:

> Language like mine cannot its force impart,
> Nor tell my sensibility of heart;
> Therefore unequal I the talk resign,
> 'Twas hers to teach, to imitate be mine.[7]

She positions herself as a student of Rowe's poetic power, although unequal in her abilities. Nevertheless, in the many lines of her poetry, Williams maps the contours of her own existence as an enquiring mind. The absence of the chief subject of her letters to the Society for the Encouragement of Arts, Manufactures and Commerce only serves to emphasise the breadth of this postmistress's interests. Williams's writing returns repeatedly to the themes of female enquiry, the wonders of nature and the human condition. Her poems offer a rebuke to those who would seek to limit the possibilities of curiosity and praise writers who open new territory for the life of the mind. Her example inspires the question: who was imitating whom in eighteenth-century society?

Ann Williams may have seen herself as an imitator of others, but her letters to the Society articulate an alternative perspective.[8] She offers historians an unusually detailed glimpse of the way science operated in the home for curious individuals who did not possess the traditional hallmarks of scholarly identity. In doing so, she revealed not only her own rigorous and encompassing enquiries into natural knowledge, but also the spatial, material and emotional dimensions of that project. Ann Williams realised and described an eighteenth-century culture of curiosity that has traditionally fallen below the radar, a culture embedded in the everyday and the home. Her example suggests that scientific methods and ideas were initiated, rather than imitated, by a wide variety of people in domestic surroundings.

There are very good reasons why it has been difficult for historians to identify, locate and examine this kind of knowledge-making and its protagonists, in all their diversity and obscurity. It hardly needs repeating that a dependence on textual survival is detrimental to the discovery of non-elite and non-textual occurrences. As discussed in Chapter 2, tacit knowledge of the kind developed at home and of crucial importance to science is not only rarely written down, it is genuinely difficult to write down. For many domestic workers, with little formal literacy, it was exactly this kind of expertise that guided their daily work. It was in the unwritten, often unspoken, exercise of technique that their specialism lay. Of course, oral cultures are commonplace in all societies, but they are of special importance where writing things down is not the primary objective of a given role. Conversation, demonstration and verbal instruction were clearly the most usual methods by which knowledge and skill were transferred person to person, especially for those who laboured by hand and, certainly, for most forms of domestic work. Between the demands of tacit knowledge and the conditions of partial literacy, many kinds of knowledge-making were rarely captured in ink or print. On this basis, it seems possible to conclude that the collection of textual fragments that comprise this volume hint at a much larger picture.[9]

In stark contrast to the silences around oral culture and tacit knowledge, a great deal has been written about writing. Unsurprisingly, this is yet another realm riddled with intellectual hierarchies that obscure the value and influence of some textual forms. A canon of published works sits atop a pyramid comprised of lesser writers, the periodical press, cheap print and ephemera. Manuscript survival nestles somewhere underneath in a period primarily known for its exponentially expanding print culture. Nevertheless, the record-keeping of house and home represents a vast body of work, although one largely devoid of literary merit. Whilst the majority of these eclectic materials were functional in purpose, evocative glimpses of personhood are visible. Lists abound – lists of servants' wages, lists of furniture, lists of plants, lists of purchases, lists of treasured objects. This ubiquitous process of accumulation and documentation structured householders' worlds – offering the promise of control in the present, a self-expression of sorts and the prospect of an archival afterlife.

The argument made here is that the home and the individuals working within that space were integral to the development of natural knowledge, certainly in terms of its motivations, practices and techniques but also in respect of its insights. This approach owes a debt to science studies and the notion that separating science from culture is unproductive and, therefore, that a defined model of production, dissemination and consumption of scientific ideas not only obscures the realities of knowledge-making but also imposes a false division between science proper and the rest.[10] Whilst the evidence of some elite lives has formed a part of the analysis here, the focus has been on the non-elite and the far-reaching significance of their demonstrable intellectual agency. It is important to emphasise that almost everyone learned much of what they knew about the world around them from home – even Fellows of the Royal Society – and this book has made the case for the home as a site of emergence in terms of scientific knowledge in this period.

Gender, status, labour and the home

Social, cultural and economic historians have reached relative consensus in characterising the eighteenth century as a period of change, notwithstanding underlying continuities.[11] One facet of this change was an increase in the quantity and diversity of materials and material culture available to individuals across society, although more generously for some. With the availability of new or rare ingredients, including expensive imported luxuries and a proliferating variety of domestic objects, it is possible to see the expansive, experimental potential of the home.[12]

Likewise, historians of science recognise this period as one of frenetic activity, with considerable developments made in fields such as natural historical taxonomy, magnetism and electricity and chemistry.[13] The eighteenth century is particularly associated with codification and the disembedding of knowledge from 'the matrix of experience it seeks to explain', as Michael McKeon has described.[14] As submissions to institutions like the Society for the Encouragement of Arts, Manufactures and Commerce demonstrate, the drive for systematisation could not entirely eradicate the particular and the local in all their messy abundance.[15]

Whilst debate continues, the old certainties of 'Enlightenment' and 'industrial revolution', looking inevitably towards the horizon of 'modernity', are now firmly contested. For historians of all stripes, a 'trickle down' or social emulation model of change over time has lost its allure.[16]

Whilst historians share a recognition of the eighteenth century as a time of societal change, there are some realms that have resisted the sheen of the new. As Sara Pennell has argued, the eighteenth-century kitchen has traditionally been characterised as a place of relative stasis – the labour-saving technologies of the nineteenth and twentieth centuries a good distance off.[17] Contrary to the upsurge in scholarship of the 1990s and early 2000s that characterised the eighteenth-century domestic interior as a dynamic space of social, gender, material and economic relations, the productive sides of these households have retained their association with drudgery.[18]

In part, this lacuna can be explained by an overwhelming focus on the consumption of the middling sorts and new forms of commerce as drivers of change in the historiography of the turn of the twenty-first century. This preoccupation with the acquisition of domestic furnishings that conveyed good taste, fashion, comfort and personal identity in places such as the dining room, parlour or bedchamber has obscured other questions.[19] As Jane Hamlett has summarised, social and cultural research on the British domestic interior has focused its energies on nuancing histories of consumption, exploring the complexities of 'public' and 'private' in the context of home and considering power dynamics and the construction of gendered identities.[20] As Amanda Vickery puts it, studies have established the 'determining role of house and home in power and emotion, status and choices'.[21] In this way, histories of the home have ensured that gender is a fundamental category of analysis and one that illuminates power relations, household decision-making and the construction of identities.[22] All of these themes contribute to the analysis offered here, but this book has positioned enquiry as a key facet of domestic work. This research thus applies the social and cultural historian's attentiveness to material culture and gendered power dynamics to the investigation of the home as a site of enquiry.

This book also follows in what is now a several-decades-long tradition of researching women's involvement in science. In the 1990s,

Lynette Hunter and Sarah Hutton introduced their ground-breaking edited collection on early modern science and medicine with an assessment of how women were subsequently excluded from these fields of enquiry. They argued that the emergence of 'modernity' in the eighteenth and nineteenth centuries demanded a new separation of knowledge from the everyday, one manifestation being the institutionalisation of knowledge. The timeframe for these major shifts depended on the type of science in question.[23]

Whilst this is a compelling narrative, it relies heavily on the slippery concept of 'modernity' and a sense that it arrived at a particular time. The concept has framed twentieth-century understandings of what came before in powerful ways. For the arguments put forward here, the term is unhelpful – at best it ensures a reading of eighteenth-century domestic enquiry as something that was about to come to an abrupt end, with the institutionalisation of science and the development of new academic disciplines. Whilst the nineteenth century certainly saw seismic shifts in these realms, it seems likely that domestic activity in that 'modern' era was also overlooked for its intellectual potential.[24] If anything, the further development of institutions packed with affluent men signals the need to delve deeper into the spaces and places that did not fit that model and were delegitimised by its hegemony.

To this end, recent scholarship has convincingly argued for extra-institutional knowledge-making as a constant from the early modern through to the contemporary. Whilst it is worth acknowledging that Hunter and Hutton were explicitly interested in female participation in science, Donald Opitz et al. emphasise the particular role of domesticity in science from the seventeenth to the twenty-first centuries.[25] Domestic science is certainly worthy of greater consideration as a continuity over the *longue durée*, not just in terms of the household as a space in which underacknowledged and underestimated scientists worked, but also as a nursery for scientific techniques and ways of knowing. Trajectories of change over time in the field of technology, for example, often suggest that humankind moved from simple tools in the pre-modern period to complex machines in the modern era. As Timothy Ingold has argued, it might be more productive to see change over time in relation to the removal of the skilled producer from the centre to the periphery.[26] The findings outlined in this book suggest, further, that the lower-status individual was marginalised in the story of their own knowledge-making well before the development of the steam train.

A burgeoning field of scholarship has emerged on the home as a site of knowledge-making. Some studies make the case for the homes of well-known naturalists and natural philosophers, such as Elizabeth Yale's analysis of John and Margaret Ray's scientific household, which is described as 'a site of mixed-gender natural inquiry and neighbourly conversation, linked to the wider world through letters and travel'.[27] Other scholars push this approach still further by positioning family life as an integral part of knowledge-making in a process that involves 'bringing family history to bear on the history of ideas' and ensuring that the life of the mind and the life of the family are 'reconstructed together'.[28]

Contrary to the notion that the 1800s brought a greater association between women and the home, John Randolph's study of the Bakunin family and Russian idealism sees 'home life's function as a theatre of intellectual activity' and examines 'self-consciously "enlightened" noblemen' choosing 'to imagine the home itself as reason's proper forum'.[29] This example takes famous male intellectuals as the starting point, again, but enlivens the domestic as a central dynamic of their work and scholarly culture. Here, the analysis corroborates the centrality of the home in intellectual culture but emphasises instead the way that households – in all their material fullness and social complexity – provided the conditions for science. Although the model of increasingly 'separate spheres' for men and women in this period has been largely rejected, it is still true that women's agency has been routinely underestimated or outright erased and that the home was simultaneously a key site of female activity. By training the lens on the home, female action, agency and enquiry become much more fully apparent.

Whilst it is difficult to uphold categorical statements in terms of diverse populations of curious men and women in this period, some observations about gender can be made. For one, an examination of the records of the Society for the Encouragement of Arts, Manufactures and Commerce reveals that male correspondents typically obscured the context and particularity of home in their narratives of enquiry and innovation. Clearly, most of these men's experiments took place at home; had another space been available, it would likely have been mentioned to underline professional standing or access to specialised equipment. Nevertheless, the dynamics and material culture of home are rarely explicitly described. By contrast, some of the richest examples of domestic detail included in this study derived from women's letters to the Society concerning their experiments with silkworms. Of course, much has been written about women's domestic roles and, as studies have shown, large quantities of – often unacknowledged – female labour was poured into producing the necessaries for human sustenance, care and comfort.[30] Nevertheless, this distinction is telling in that it underlines the centrality of gender in shaping interactions with domestic space, personnel and equipment.

Not only were the use of space and the obligations of that space divergent for men and women, but the way hours in the day were divided and distributed differed too. Personal autonomy and the ability to choose how to spend time depended on gender, status and role. An individual's mastery over their own time and its use had a direct relationship with their own personal power and freedom to impose upon the time of others. The Dublin apprentices discussed in Chapters 4 and 6 clearly delineated the time they had before and after their day's work, because this was the time that was theirs to use as they pleased. Robert Jackson's walk between his Meath Street residence and a preferred location for astronomy was measured down to the minute – revealing his attention to these slivers of time that could afford or deny his favourite pursuit. Likewise, Ann Williams fitted her enquiries into short bursts of time distributed across the day in a home that was located at her place of work – a post office. Both Thomas Chandlee and Ann Williams, interestingly, compared their working lives to those of enslaved people. These comments reveal their ignorance of the real experiences of the enslaved but also hint at the resentment that many curious individuals felt about the constraints on their time and attention.[31]

Clearly, temporal patterns of labour depended on the specific kind of work at hand and the dramatic seasonal shifts of agriculture differed from the monthly or yearly rhythm of work in an urban workshop. The temporality of one sphere might also influence activity in another. The cycle of seasons strongly informed the calendar of religious days and these, in turn, influenced aspects of state bureaucracy such as tax collecting or the regularity of court sessions.[32] The way people's lives were structured by time owed a debt to a confluence of different factors. Furthermore, time and the way it was regulated and measured were contested in the eighteenth century. For many, the almanac set out the calendars of natural and man-made occurrences but, increasingly, people used clocks and watches to organise their days. Major calendar reform in 1752 brought its own conflicts and the uniformity and universality of clock time remained some decades in the future.[33] Thus, the way individuals like Williams, Rhodes, Jackson and Chandlee understood the organisation and use of their time varied and operated with considerable mutability as compared with subsequent centuries.

Differences in terms of domestic roles clearly also had consequences for identity and how people understood and communicated the basis for their intellectual authority, as discussed in Chapter 6. Whilst histories and geographies of science have rightly emphasised place and space as crucial facets of intellectual work in this era, and this book focuses on one such space, time is an overlooked dimension in analyses of enquiry, especially as it took place among the competing demands of home.

In one sense, this book represents an intellectual history from below. However, the findings suggest that lower-status scientists were not just ignored, but their work was also misunderstood, with consequences for how knowledge-making is characterised more generally. These findings fit models of indigenous knowledge much better than they do standard histories of western enquiry.[34] Here, people were doing things in a space that was associated with low-status work, but which was in fact highly complex and dynamic. The scientists in this book made their enquiries by responding to that space, its material and spatial facets, its social and emotional draws and its temporal patterns. Their doing and thinking were impossible to imagine without the conditions and actors (human and non-human) of that place.[35] These scientists did not pursue singular questions in ways that ignored all of their other concerns and objectives; they investigated amidst and *through* the myriad of materials, tasks and schedules that were inherent to their place of enquiry. They did their work with their environment, not in spite of it.

The everyday

This book has placed the home at the heart of the action in terms of scientific practices and understanding, precisely because of its foundational role in social relations. The research thereby replaces a framework of centre and periphery with 'patterns of mutual interdependence' and responds to James Secord's call to see science as 'a form of communicative action'.[36] Here, this communication takes place between people and things.[37] The research has aimed to unsettle dichotomous categories by focusing on what people did at home, day to day and seeing the full range of activities as worthy of attention. This approach draws on Ingold's understanding of a person as 'a singular locus of creative growth within a continually unfolding field of relationships'.[38] His rejection of the separation of fields of enquiry leads to a proposal in favour of looking at 'skilled practices of socially situated agents' to better understand ways of knowing.[39] This book attempts to do just that for eighteenth-century knowledge-making.

The analysis also takes inspiration from Michel de Certeau's articulation of the 'clandestine forms taken by the dispersed, tactical, and make-shift creativity of groups or individuals'.[40] Of course, the activity of the obscure scientists discussed here might appear more clandestine than it perhaps was because of how subsequent generations have taken to preserving, understanding and explaining the past. Nevertheless, science as conducted at home was often dispersed, tactical and makeshift in character, whether it was conducted by famous scientists or others.[41] The extent to which this science of the home should be mainly viewed as the unacknowledged work of the many (claimed by the few) or the subversive activity of the marginalised, leveraging daily practice to make their mark on the socioeconomic regime, remains to be seen. What is very clear is that science was embedded in practices of (family) life and work and inseparable from them and, as Lorraine Daston has stated, there is no way 'of excising science cleanly from other ways of knowing and doing'.[42] In this appraisal, all knowledge-making is 'everyday', perhaps relieving historians of the responsibility to demarcate it as such.

For now, however, it remains important to describe and explain the everyday quality of knowledge-making because academic disciplines are still shaped by hierarchies and dichotomies that serve to divide the cerebral from the rest. Whilst scholars have recognised the fundamental importance of the everyday, its precise characteristics in varied contexts are worthy of further analysis. If knowledge-making emerges from the everyday, then how does it do so and what difference does that make? In some ways, the shift in thinking here seems to rest on the re-categorisation of the rare and exclusive as mundane and accessible. However, Christel Avendal has identified the concept of 'heightened everydayness' or 'the extraordinariness in the ordinary' as it relates to the 'unnoticed knowledge' of daily existence.[43] Chandra Mukerji takes the argument further. Drawing on Pierre Bourdieu's concept of 'habitus', she suggests that these socialised norms of everyday existence 'may silently reproduce relations of power most of the time, but ... can also turn trickster, using tacit knowledge to pursue dreams and hone aspirations'.[44] When men wrote to the Society for the Encouragement of Arts, Manufactures and Commerce about their home experiments without explicitly mentioning the home, they attempted to disguise the mundane in the telling of the extraordinary. They considered the everyday ill-suited to the new and the highly prized. Nonetheless, it was out of these circumstances and through practice that leaps of imagination materialised.

In demonstrating the lack of any clear division between activities that are categorised as mundane and those that are considered to be exploratory, this book deliberately blurs distinctions between extraordinary and ordinary. Whilst specific strands of domestic knowledge-making have been examined by scholars, the full spectrum of interactions between the quotidian and the enquiring is underexplored.[45] This book has captured examples of domestically situated knowledge-making that emerge from the particulars of that environment, rather than being imposed upon it.

One way to grasp the complexity of the everyday is to focus attention on specific practices within a larger ecology of activity, whether that is in the home or another multi-purpose space. Some practices, such as producing and consuming sustenance, are intrinsically linked with the home, whereas others are much more loosely associated with that location. This book began with a discussion of practices, such as cooking, that are inseparable from home and moved through to practices that have explicitly scientific connotations. As others have shown, science was conducted at home by the most famous scientists of this era; nonetheless, most histories of the household have little to say about observation and experiment. Collecting sits somewhere in between, strongly associated with the homes of a few affluent and dedicated collectors, such as Hans Sloane or the Duchess of Portland, but much less clearly related to the homes of ordinary people.

The practice of collecting points to an important issue across the board – the question of extent. Clearly, a person could collect a few artefacts or specimens without being considered a 'collector' in the rather grander sense of the word. Similarly, someone could record many details about household expenses without being the kind of record-keeper who attended to the weather, the stars or their local flora and fauna. Observation is entirely invisible to the historian's eye if the observer had no urge to record the details of what they saw. The same is true for experiment, and the lengths an enquiring mind went to in this regard varied widely. However, if scholarly understanding of science is structured around a threshold of achievement (however that is chosen) it will result in the exclusion of those individuals who did not make that pre-determined level or did not leave clear evidence that they did so. This approach also individualises the issue, driving the analysis towards particular, extraordinary people rather than the culture that enabled their enquiry. By switching attention to the context of enquiry, its enabling and disabling characteristics and the actual things that people did day to day, a much more interesting scene is visible.

The language of enquiry

Despite major developments in the definition of what counts as science, the urge to measure agency and action against something dichotomously powerless and inactive is still strong. As such, language – as many theorists have described – actively limits the shape and scope of a given enquiry, not least concerning the subject of this book.[46] As Roger Cooter and Stephen Pumfrey notably argued, 'a binary logic' pervaded twentieth-century historical accounts of early modern science, reducing the terms of debate to rational science versus magic and superstition with the attendant ramifications for narratives of change over time.[47] In general terms, binaries are almost never comprised of equal partners and thinking in dualistic terms not only diminishes one-half of the pair but distorts both parties.[48] Of course, many scholars have analysed this issue, but it is worth naming some of these troublesome binaries and considering how to move beyond them.

In terms of the question of knowledge-making, hand and mind is the over-arching and ancient dichotomy and the hand has certainly been the poor relation to the mind in traditional accounts of learning. Of course, that privileging of mind over body has extremely long roots and the question has been dealt with extensively by historians of science, especially those who examine the changes taking place in the early modern period. As a result of this rich vein of scholarship, the influence of artisanal knowledge in the history of science is now taken as read. Besides the scholar/artisan dichotomy, other combinations can be equally unhelpful, for example, science/technology, pure/applied, theory/practice, experimental/mathematical or art/nature.[49] The list could be extended still further and the influence of the structures these dichotomies create is difficult to over-estimate.

In the process of researching this book, it has been a struggle to find language to describe the curious people visible in the archive without falling into the same kind of binary formulation, naturally denigrating one-half of the pair. For example, the individuals discussed here worked at home and largely outside of the membership of learned societies. The terms *informal* and *formal* might offer a way of describing the difference between the two, but the knowledge-making of a housewife does not seem especially informal as compared to the knowledge-making of a Fellow of the Royal Society – usually a leisured man with significant resources at his disposal.[50] Her knowledge was created within a knowable context while undertaking productive labour, it required time and dedication to bring about, it draws upon other widely accepted laws, rules or findings and it holds the potential to contribute to the shared well of human understanding. There is another discussion to be had about the influence this housewife's work might have had on the print culture of eighteenth-century science, or the purchase it may or may not have had on the debates of well-known scholars, but there doesn't seem to be anything intrinsically *informal* about what she does.

Further, in-depth discussion of observation and experiment usually relates to natural philosophers. In this context, these practices are described as narrow and specific; the scientific observer 'trains and strains the senses, molds the body to unnatural postures, taxes patience, focuses … attention on a few chosen objects at the expense of all others'.[51] These actions sound atypical; the distinction rests on the way these forms of attention contrast with the socially acceptable gestures of the rest of life. However, the unnatural pose of the brewer, atop a ladder, intent on the process of fermentation does not seem entirely dissimilar to the naturalist squinting at their object of interest – likewise for the cook's precise, swift and responsive actions. The findings of this book suggest an interpretation of practices of attention as on a spectrum that includes the naturalist and the brewer; the chemist and the cook.

As Jane Whittle has argued in the field of pre-industrial economic history, work conducted at home was only given the diminishing prefix of 'domestic' when it was done by women and was, therefore, erroneously considered to be focused on care or subsistence and of low importance to the market economy. Male work done at home was categorised differently (often as agriculture or construction) regardless of its relationship with subsistence and deemed of intrinsic relevance to the wider economy.[52] Here the distorting effects of the epithet 'domestic' and its gendered associations become abundantly clear. This is no less true for the domestic labour of enquiry, integrated as it was with the sustaining, productive and economic facets of household work.

Beyond descriptions of the people who enquired into nature's secrets, descriptions of knowledge itself are similarly affected by dichotomous categories. There are a variety of common formulations – all of which seem to offer an alternative to what is usually called scholarship. This supposedly formal, professional, educated, elite scholarship is not often fully defined, but for those who are interested in looking at different kinds of intellectual actors, the urge to caveat the knowledge they have with words like 'everyday' or 'useful' is strong. Sometimes the term 'knowledge' is simply exchanged for 'know-how', suggesting that someone possesses skill and technique without understanding why they work. This was certainly the way that Bishop Edward Synge characterised the knowledge of his servant, Jane, in Chapter 2. It is unsurprising that he did so.

Clearly, eighteenth-century society also made distinctions between knowledge that helped people understand the workings of nature and knowledge that was useful, especially in terms of economic benefit. The Society for the Encouragement of Arts, Manufactures and Commerce and the Dublin Society are key examples of this period's interest in 'applied knowledge' as it is called today. Despite that distinction, this question of language and esteem remains, especially if twenty-first-century historians are to avoid the distorting influence of age-old intellectual hierarchies.[53] Sarah Easterby-Smith has stressed the importance of moving beyond hegemonic frameworks for understanding knowledge-making by using micro-historical approaches, but acknowledges the gravitational pull of 'concepts based in western knowledge systems that might occlude other forms of knowing'.[54] The problem here is not that there is a lack of critical appraisal of these dichotomies and their influence on conceptions of knowledge-making but that, despite scholars' protestations, histories still lean on them – absent-mindedly – precisely because understandings of the world, then and now, are so rooted in binary formulations.[55]

Recent scholarship has used the prefixes 'everyday' and 'useful' to characterise the distinctive conditions and personnel of some knowledge-making.[56] The knowledge discussed here is not considered to be distinctively or exclusively useful or everyday, or at least no more so than the knowledge that emerged from other spaces in this period. Whyman, Leong and others' articulation of specific cultural and class contexts for knowledge-making contributes greatly to diversity, regionality and specificity in histories of knowledge.[57] This developing plurality of knowledges and knowledge-makers is another important step in resisting false dichotomies and the flattening effect of mono-cultural explanations. Whilst the research presented here contributes to this project, the decision not to prefix the knowledges and knowledge-making described is a deliberate one. Taking inspiration from feminist scholars' efforts to redirect attention away from discussions that emphasise 'women's issues' or a female sphere, this book positions its discussion as part of the 'mainstream', thereby underlining the centrality of domestic activity in histories of knowledge, society, culture or the economy.[58]

All knowledge-making is essentially a product of the 'everyday' and to describe the home as especially quotidian in character wrongly implies that it is inherently more habitual and routine than other environments. Furthermore, the ordinariness of this environment seems inevitably and dichotomously contrasted with the extraordinary potential of other spaces. Whilst it has not been possible to pursue a deep comparison between the household and an alternative site of knowledge-making to assess their relative qualities, this book argues that the remarkable nestles, at ease, in the contours of the everyday, domestic or otherwise.

As the ecofeminist and philosopher Val Plumwood has argued, there are a number of routes out of dualism.[59] To begin, a thorough examination of the less valued half of the binary construction is needed – one that does not ask questions structured by dualist assumption. For example, just as this book has argued that Fellows of the Royal Society are not the only scientists worthy of notice, it does not contend that lower-status, domestic experimenters were the only 'real' intellectuals at work in this period, thereby inverting the binary power dynamic. As Plumwood advises, this book has understood the dualism of mind and hand as serving to distort the features of both mind and hand or, equally, man and woman; master and servant; coloniser and colonised.[60] By enacting this distortion, the dualism has overlooked the domestically rooted nature of science, insisting on seeing the elite, male scientist's intellectual work as both remarkable and in direct contradistinction to everyday work and the 'female' spaces, skill and knowledge through which it was conducted. However, whilst describing the less well described, this research has also included the wealthy and the privileged in order to see the similarities and disparities that emerge and to make visible the connections between making bread and collecting expensive objects. By affirming the unacknowledged, redefining it in relation to the hegemonic and reconstructing a sense of the whole, dualistic thinking can be circumvented. This approach aims to upend some core assumptions about enquiry concerning where it took place, who did it and the actions that catalysed leaps in thinking in this period – thinking that, of course, was borne out of human and non-human collaboration.

From individual knowing to societal knowing; or, knowledge trickles upwards

When 'attention is paid to gender and geographies, and when hierarchies of knowledge production are rejected' a different kind of 'knowledge society' emerges.[61] However, detractors might caution against the danger of attributing 'the same status to the growing of cucumbers as to the practice of particle physics'.[62] Clearly, for many of those engaged with the history of knowledge, the knowledge itself – its quality and character – will always matter. Whilst cucumbers are admittedly a subject covered, briefly, by this book, the objective here has been to illuminate the conditions that generated knowledge-making rather than assess the products of that process.

The qualitative detail available about the activities of curious individuals is revealing as to the communicative nature of their scientific practices. In those cases where there are multiple letters describing engagements with science, it becomes obvious that men and women commonly operated as part of communities of learning – communities that encompassed their family, friends, fellow workers and neighbours but extended to communities forged through letter-writing, print culture and reading. In these ways, the curious operated at several levels and scales – perhaps rooted in a particular location but moving through diffuse networks of association and information sharing. Institutions and the periodical press were instrumental in promoting these networks of connection between interested individuals and the possibility of exchange, but – of course – people also developed their own social and intellectual communities independently of such infrastructure, correspondence being a powerful technology of the curious.[63] Moreover, the significance of many institutions and publications could be attributed more fairly to the hordes of willing contributors than the founders and editors of such outlets. Without a highly willing public, eighteenth-century print culture would not have generated the volume of lively exchange for which it is now famed.

In addition, as Chapter 2 illuminated, oral culture must have been even more important to localised information-sharing and the acquisition of technique and tacit knowledge than text. When Bishop Synge took pen in hand to describe the process of learning how to bake bread from his servant, Jane, he enacted an unusual thing – the textual articulation of learning through doing. His prose struggled with that task as words find embodied, material practices particularly hard to wrap themselves around. Moreover, his account revealed the different words that a landed gentleman and a domestic servant used to describe material processes. Each party had an oral culture, but they differed in distinct ways. Synge's letters document these disparities and gaps in language and understanding as he attempted to concretise and codify the tacit knowledge of home provisioning. However faulty the relaying of tacit knowledge in textual form, this evidence provides a glimpse of a crucial layer of knowledge-making and sharing in eighteenth-century society. When it came to manipulating materials to make useful products, oral culture was the medium for learning and improving. Whilst this book relies on textual survivals, it aims to acknowledge this largely invisible layer of scientific exchange. Oral culture provided a large part of the talkativeness of scientific activity in this period and contributes to the view of enquiry as an inherently social pursuit.

Silkworm experimenters, Williams and Rhodes, found people within their locale to share their interests and admire their work, but likewise courted attention from people further afield who had heard of their endeavours.[64] Both women engaged with a range of printed material on their favoured subject, perhaps directly from the Society's *Transactions* but likely also as reproduced cheaply in the periodical press. Similarly, Dublin apprentices Jackson and Chandlee marshalled a city-based community of star-gazers who they met with in person in public spaces such as St Stephen's Green. However, whilst they established a traffic of exchange between, mainly, working men of this city, Jackson and Chandlee found a much wider horizon through the coverage of astronomy in magazines and almanacs. They were masters of the genre – using Jackson's access through the print trade and insider knowledge to trawl a great variety of domestic and imported publications, comparing, contrasting and critiquing the information contained. For men whose best hours of the day were absorbed with the demands of an exacting master, their communicative and communal approach served them well.

By taking the home as a vantage point, it is still possible to see the gravitational pull of institutions, as they actively enticed individuals from across the world to share their observations and describe their experiments and to do so in conversation with one another. Conversely, those who wrote into various societies, or for that matter the *Gentleman's Magazine*, were also engaged in other profound and demanding networks of association: family, neighbourhood and occupation.[65] Ann Williams's foray into the world of publication precluded any mention of silkworms, although – of course – her lines of ink on that subject were subsequently reproduced in print by the Society for the Encouragement of Arts, Manufactures and Commerce. It would seem extremely likely that an 'A. W.' or 'A postmistress, Gravesend' penned an occasional letter to a magazine about a subject of public interest. In these ways, Williams operated in multiple 'spaces' of exchange but her contributions were based on the bedrock of her experience gained through the conditions of her labour and life. The people discussed here may occasionally have been eccentric, but they were not unusual in terms of the activities they engaged in. They just happened to be the ones who put their words onto the page and whose pages were fortunate enough to be preserved.

Conclusion

Homes were spaces of emergence, in terms of the drivers, practices and techniques of science, but they were also places where insight and innovation could gain traction. The variegated demands, affordances and schedules of domestic and familial life provided not only the conditions for enquiry but also, often, the motivation. Intellectual labour was just one facet of household work and scientific activity that operated within and around the rhythms of provisioning, oeconomy, sociability and care. To recognise scientific enquiry as household labour alters its complexion, but this book has sought to demonstrate the historical insight made possible by that shift. The research recognises that other spaces were also crucial to the development of natural knowledge but has focused on a place that was not only multifunctional, but also freighted in terms of its meanings in ways that have served to obscure its intellectual significance. Here, the eighteenth-century home is viewed as a dynamic, creative, communicative and connected place and the analysis has striven to escape the diminishing connotations of 'domestic', based as they are on hierarchies that historians have long rejected.

Most scientific projects were necessarily collective endeavours, relying on the collation and comparison of data from a number of quarters. The examples explored here also point towards the synergies between processes of categorising and organising present in both natural history and home 'oeconomy' – the search for order amidst the clutter of accumulation and the trial and error of experiment.[66] Many of the curious people discussed in this book saw their endeavours in this light, as contributing a small part to the growing bedrock of human understanding. They often considered it natural for the British, wherever they were across the globe, to take the lead in knowledge-making and to apply new insights to further national and imperial objectives. Where manipulation of the natural world was perceived as more successful abroad, steps were taken to acquire and apply that knowledge to advantage at home.

The observers and experimenters that have left the deepest mark on this book are those who wrote eagerly to one another or an institution to report their findings, in the process explaining in lively detail the content and meaning of their endeavours. Some of the more voluble individuals were those who worked for their living, making their commitment to enquiry seemingly remarkable as compared with those leisured few who had the freedom to make it their vocation. However, the analysis here has sought to dislodge an oversimplified association between curiosity and leisure. By focusing on the generative conditions of the home, the domestic worker's understanding can be seen as an advantage rather than a detriment. In the gentry's efforts to appropriate such knowledge, the very real value it held for them at this time becomes clear.[67] Collectively, they represent a deep-rooted culture of curiosity. The curiosity of these people forged and animated the search for natural knowledge in this period and they quietly conducted this complex project from the comfort of their very many different homes.

Notes

1 Williams, *Original poems*, frontispiece; alongside Thynne (later first Baron Cartaret, 1735–1826) there is a co-dedication to Francis Dashwood, Eleventh Baron Le Despencer (1708–81) who was the second Post-Master General from 1766 until his death; for more on Williams see Chapters 5 and 6.

2 General James Wolfe (1727–59) was a British army officer who is primarily remembered for his 1759 defeat of the French in Quebec.

3 Williams, *Original poems*, pp. 72, 104.

4 *Ibid.*, pp. 111, 149, 163.
5 *Ibid.*, pp. 58, 190–1; referring to the historian Catharine Macaulay (1731–91) and the poet Elizabeth Singer Rowe (1674–1737).
6 Williams, *Original poems*, p. 190; it is worth noting that most girls were educated at home in this period – an education that has traditionally been characterised as inferior to formal schooling but which was often full and diverse in its content; see Cohen, 'To think, to compare, to combine'.
7 Williams, *Original poems*, p. 191; literary imitation was a common practice in this period, see Robert L. Mack, *The genius of parody: Imitation and originality in seventeenth- and eighteenth-century English literature* (Basingstoke: Palgrave Macmillan, 2007).
8 Discussed in detail in Chapters 5 and 6.
9 See, for example, the Bishop Synge letters in Chapter 2 – an unusual exploration of tacit knowledge, authority, gender relations and the struggle to write what you do.
10 See, for example, Richard Whitley, 'Knowledge producers and knowledge acquirers: Popularisation as a relation between scientific fields and their publics' in Terry Shinn and Richard Whitley (eds), *Expository science: Forms and functions of popularisation* (Dordrecht: D. Reidel, 1985), pp. 3–28; see also Secord, 'Knowledge in transit'.
11 For studies that consider the period as one of significant change, see: Jeff Horn, Leonard N. Rosenband and Merritt Roe Smith, *Reconceptualizing the industrial revolution* (Cambridge, MA: MIT Press, 2010); Joel Mokyr, *The gifts of Athena: Historical origins of the knowledge economy* (Princeton, NJ: Princeton University Press, 2002); Jan de Vries, *Consumer behavior and the household economy, 1650 to the present* (Cambridge: Cambridge University Press, 2008); E. A. Wrigley, *Energy and the English industrial revolution* (Cambridge: Cambridge University Press, 2010); see also Lissa Roberts and Simon Werrett on material and knowledge production as an axis for societal change in late eighteenth and early nineteenth centuries: 'Introduction: "A more intimate acquaintance"' in Roberts and Werrett, *Compound histories*, pp. 1–32, esp. p. 30.
12 Classic texts include John Brewer and Roy Porter (eds), *Consumption and the world of goods* (London: Routledge, 1994); Weatherill, *Consumer behaviour*; and Maxine Berg, *Luxury & pleasure in eighteenth-century Britain* (Oxford: Oxford University Press, 2005); see also Roy Porter, 'English society in the eighteenth century revisited' in Jeremy Black (ed.), *British politics and society from Walpole to Pitt 1742–89* (Basingstoke: Macmillan, 1990), pp. 29–52; Martin J. Daunton, *Progress and poverty: An economic and social history of*

Britain, 1700–1850 (Oxford: Oxford University Press, 1995); John Rule, *Albion's people: English society, 1714–1815* (London: Longman, 1992).

13 See, for example, Susannah Gibson, *Animal, vegetable, mineral?: How eighteenth-century science disrupted the natural order* (Oxford: Oxford University Press, 2015); Patricia Fara, *Sympathetic attractions: Magnetic practices, beliefs, and symbolism in eighteenth-century England* (Princeton, NJ: Princeton University Press, 1996); Paolo Bertucci, 'Sparks in the dark: The attraction of electricity in the eighteenth century', *Endeavour*, 31:3 (2007), pp. 88–93; Ursula Klein and Wolfgang Lefèvre, *Materials in eighteenth-century science: A historical ontology* (London: MIT, 2007).

14 Michael McKeon, *The secret history of domesticity: Public, private, and the division of knowledge* (Baltimore, MD: Johns Hopkins University Press, 2005), p. xix.

15 See Paskins, 'Sentimental industry', on disembedding practices, especially on the local specificity of agricultural development, pp. 32–40.

16 For history of science see, for example, Secord, 'Knowledge in transit'; Jonathan R. Topham, 'Focus: Historicizing popular science, introduction', *Isis*, 100:2 (2009), pp. 310–68; Lorraine Daston, 'The history of science and the history of knowledge', *KNOW: A Journal on the Formation of Knowledge*, 1:1 (2017), pp. 131–54; and for social and cultural history see, for example, Trentmann, *Empire of things*; Vickery, *Behind closed doors*, p. 304; Jonathan White, 'Review essay: a world of goods? The "consumption turn" and eighteenth-century British history', *Cultural and Social History*, 3:1 (2006), pp. 93–104.

17 Pennell, *English kitchen*.

18 With notable exceptions, especially, Pennell, *English kitchen*; Steedman, *Everyday life*.

19 See, for example, Brewer and Porter, *Consumption*; Michael Snodin and John Styles (eds), *Design and the decorative arts, Britain 1500–1900* (London: V&A Publications, 2001); Jeremy Aynsley and Charlotte Grant (eds), *Imagined interiors: Representing the domestic interior since the Renaissance* (London: V&A Publications, 2006); John Styles and Amanda Vickery (eds), *Gender, taste, and material culture in Britain and North America, 1700–1830* (New Haven, CT: Yale Center for British Art, 2006); Sasha Handley, 'Objects, emotions and an early modern bed-sheet', *History Workshop Journal*, 85 (2018), pp. 169–94.

20 Jane Hamlett, 'The British domestic interior and social and cultural history', *Cultural and Social History*, 6:1 (2009), p. 97 (pp. 97–107).

21 Vickery, *Behind closed doors*, p. 3.

22 Hamlett, 'British domestic interior', p. 97.

23 Hunter and Hutton, *Women, science and medicine*, pp. xviii–xix.
24 Opitz et al., *Domesticity*.
25 *Ibid.*; also see Alix Cooper, 'Afterword' in Opitz et al., *Domesticity*, which discusses how extra-institutional knowledge-making also stretched back into ancient times, pp. 281–7.
26 Timothy Ingold, *The perception of the environment: Essays in livelihood, dwelling and skill* (London: Routledge, 2000), pp. 294–311.
27 Elizabeth Yale, 'A letter is a paper house: Home, family, and natural knowledge' in Cala Bittel, Elaine Leong and Christine von Oertzen (eds), *Working with paper: Gendered practices in the history of knowledge* (Pittsburgh, PA: University of Pittsburgh Press, 2019), p. 159 (pp. 145–59).
28 Coen, 'Common world', pp. 428, 421; the former quotation referring to Emma Rothschild, *The inner life of empires: An eighteenth-century history* (Princeton, NJ: Princeton University Press, 2011).
29 John Randolph, *The house in the garden: The Bakunin family and the romance of Russian idealism* (London: Cornell University Press, 2007), pp. 3, 64, 65.
30 See Jane Whittle and Mark Hailwood, 'The gender division of labour in early modern England', *The Economic History Review*, 73:1 (2020), pp. 3–32; Whittle, 'Critique'; Jane Humphries, 'The wages of women in England, 1260–1850', *The Journal of Economic History*, 75:2 (2015), pp. 405–47.
31 See RSA, PR/GE/118/11/939, see also RSA, PR/GE/118/11/937–948 and FHLD, Fennell, MSS Box 27, folder 1, letter 19: Robert Jackson to Thomas Chandlee, 9 Apr. 1769.
32 David Fleming, 'Cycles, seasons and the everyday in mid-eighteenth-century provincial Ireland' in Raymond Gillespie and R. F. Foster (eds), *Irish provincial cultures in the long eighteenth century: Making the middle sort, essays for Toby Barnard* (Dublin: Four Courts Press, 2012), pp. 133–4 (pp. 133–54).
33 See, for example, Edward P. Thompson, 'Time, work-discipline and industrial-capitalism', *Past & Present*, 38 (1967), pp. 56–97; Robert Poole, '"Give us our eleven days!": Calendar reform in eighteenth-century England', *Past & Present*, 149 (1995), pp. 95–139.
34 See J. Mistry, 'Indigenous knowledges' in Rob Kitchin and Nigel Thrift (eds), *International encyclopaedia of human geography* (London: Elsevier Science, 2009), pp. 371–6; Margaret M. Bruchac, 'Indigenous knowledge and traditional knowledge' in Claire Smith (ed.), *Encyclopedia of global archaeology* (New York: Springer Science and Business Media, 2014), pp. 3814–24; Maria Franco Trindade Medeiros, *Historical ethnobiology* (Amsterdam: Academic Press, 2020), esp.

chapter 7, 'Thinking about the conceptualizations of types of knowledge and human communities', pp. 139–70.

35 See Hutchins, *Cognition in the wild.*

36 Secord, 'Knowledge in transit', pp. 669, 663; see also Alan Lester's discussion of the framework of centre and periphery in 'Spatial concepts and the historical geographies of British colonialism' in Andrew S. Thompson (ed.), *Writing imperial histories* (Manchester: Manchester University Press, 2013), pp. 118–42.

37 Bruno Latour, *Reassembling the social: An introduction to actor-network-theory* (Oxford: Oxford University Press, 2005).

38 Ingold, *Perception of the environment*, pp. 4–5.

39 *Ibid.*, p. 289.

40 Secord, 'Knowledge in transit', p. 669; de Certeau, *Practice*, p. xiv.

41 Werrett, *Thrifty science.*

42 Daston, 'The history of science', p. 145.

43 Christel Avendal, 'Heightened everydayness: Young people in rural Sweden doing everyday life' (PhD thesis, Lund University, 2021); see also www.newhistoryofknowledge.com/2021/08/27/everyday-knowledge-as-unnoticed-knowledge/ (accessed 18 February 2022).

44 Chandra Mukerji, 'The cultural power of tacit knowledge: Inarticulacy and Bourdieu's habitus', *American Journal of Cultural Sociology*, 2:3 (2014), p. 371 (pp. 348–75).

45 A good example of such a strand is the domestic knowledge associated with recipe books and its relationship with natural knowledge; see Leong, *Recipes.*

46 See, for example, Ferdinand de Saussure, *Cours de linguistique générale* (Paris: Payot, 1916); Saul Kripke, *Naming and necessity* (Cambridge, MA: Harvard University Press, 1980); and also the work of post-structuralist thinkers such as Judith Butler, Julia Kristeva, Gilles Deleuze and Michel Foucault among others.

47 Roger Cooter and Stephen Pumfrey, 'Separate spheres and public places: Reflections on the history of science popularization and science in popular culture', *History of Science*, 32:3 (1994), p. 240 (pp. 237–67); Lorraine Daston has argued for the history of science to rename itself the history of knowledge precisely because of the intractable relationship of the former with erroneous notions of progress, modernity and western hegemony; see 'The history of science'.

48 Val Plumwood, *Feminism and the mastery of nature* (London: Routledge, 1993), esp. chapter 2, pp. 41–68.

49 Roberts et al., *Mindful hand*, p. xiv; Long, *Artisan/practitioners*, p. 7.

50 See Michèle Cohen's discussion of the way girls' education was routinely described as 'informal', 'unsystematic' and 'superficial' in contrast

with that of their brothers in '"Familiar conversation": The role of the "familiar format" in education in eighteenth- and nineteenth-century England' in Jill Shefrin and Mary Hilton (eds), *Educating the child in enlightenment Britain: Beliefs, cultures, practices* (London: Routledge, 2009), pp. 99–117.

51 Daston and Galison, *Objectivity*, p. 234.

52 Whittle, 'Critique'; her conclusions have far-reaching consequences that reach far beyond pre-industrial periods and into the present day.

53 See, for example, Carroll, 'Politics of "originality"'.

54 Easterby-Smith, 'Recalcitrant seeds', p. 222; for a critique of micro-history as the primary approach see Peter Galison, 'Limits of localism: The scale of sight' in Wendy Doniger, Peter Galison and Susan Neiman (eds), *What reason promises: Essays on reason, nature and history* (Berlin: De Gruyter, 2016), pp. 155–70.

55 Examples include Elizabeth Yale's introductory lines to her volume *Sociable knowledge: Natural history and the nation in early modern Britain* (Philadelphia, PA: University of Pennsylvania Press, 2016), p. 1, where naturalists and antiquaries 'collaboratively constructed their visions ... and through their printed works, they communicated these visions to a wider public', over-playing the division between knowledge-makers and a passive 'wider public'.

56 Leong, *Recipes*, p. 6; Whyman, *Useful knowledge*, esp. pp. 4, 13; Pamela H. Smith also discusses 'vernacular science' in 'Vermilion, mercury, blood, and lizards: Matter and meaning in metalwork' in Ursula Klein and Emma C. Spary (eds), *Materials and expertise in early modern Europe: Between market and laboratory* (Chicago, IL: Chicago University Press, 2010), pp. 41, 47–8 (pp. 29–49).

57 See also Secord, 'Science in the pub'; Anne Secord, 'Elizabeth Gaskell and the artisan naturalists of Manchester', *The Gaskell Society Journal*, 19 (2005), pp. 34–51.

58 See, for example, Maria Puig de la Bellacasa, *Matters of care: Speculative ethics in more than human worlds* (Minneapolis, MN: University of Minnesota Press, 2017), who is herself indebted to the feminist political philosophy of Joan Tronto, *Moral boundaries: A political argument for an ethic of care* (New York: Routledge, 1992).

59 Plumwood, *Feminism*, pp. 59–68.

60 *Ibid.*, pp. 41–55.

61 von Oertzen et al., 'Finding science in surprising places', p. 74.

62 Cooter and Pumfrey, 'Separate spheres', p. 254.

63 See Leonie Hannan, *Women of letters: Gender, writing and the life of the mind in early modern England* (Manchester: Manchester University Press, 2016).

64 See, for example, RSA, *Transactions*, vol. 2 (1784), p. 161 and the discussion in Chapter 6.

65 On the importance of considering emotion and the family in relation to science, see Coen, 'Common world'.

66 See Jonathan Sheehan and Dror Wahrman, *Invisible hands: Self-organisation and the eighteenth century* (Chicago, IL: University of Chicago Press, 2015) for a broader philosophical and cultural exploration of this concept and its emergence in this period.

67 Fisher, 'Master should know more'; Peter M. Jones, *Agricultural enlightenment: Knowledge, technology, and nature, 1750–1840* (Oxford: Oxford University Press, 2016).

Bibliography

Primary sources

Manuscript

British Library

Clive Papers, Mss Eur Photo Eur 287.
Evelyn Papers, Add MS 78539.
Powys Diaries, Add MS 42173.
Trumbull Papers, Add MS 72516.

Derbyshire Record Office

Wright of Eyam Hall, D5430/50/4–5.

Dublin City Library and Archive

Gilbert Collection, MS 132, Isaac Butler, 'A Diary of Weather and Winds', 1716–34.

Dublin Society

Minute Book, vol. 6, 9 Mar. 1758 to 13 Aug. 1761.

East Sussex Record Office

SAS/CP 293: Eastbourne Place Inventory.

Friends Historical Library Dublin

Selina Fennell Collection, MSS Box 27/Folders 1–3.

Hampshire Record Office

9M73/G847–849, catalogues and inventories.

The Irish Architectural Archive

2/94 CS4, Strokestown Park, Basement Still Room.

John Rylands Library

Stamford Papers, 'Household consumption account book', GB 133 EGR7/1/1–13.
Stamford Papers, 'Garden account books', 1778–1822, GB 133 EGR7/7/1–4.
Stamford Papers, 'Household inventory, 1819', GB 133 EGR7/17/3.

The National Archives

Prob 11/1198.
Prob 11/1602.

National Library of Ireland

MS 2178: Accounts of household and personal expenses of Jane Creighton, 1st Baroness Erne, 1776–1799.
MS 34,952: 'Mrs. A. W. Baker's cookery book, vol. 1, 1810'.
MS 36,375/3: O'Hara of Annaghmore Papers.
MS 42,007: 'Household account book, 1797–1832'.

Petworth House Archive

MS 2236 'House book'.
MS 8060.
MS 8065 'Bills paid'.

Royal Dublin Society

Dublin Society Minute Book, vol. 6 (9 Mar. 1758–13 Aug. 1761).

Royal Society of Arts Archive

PR/MC/101/10/468: 26 May 1791.
PR/MC/105/10/469: 21 Apr. 1799.
PR/MC/105/10/470: 20 Jul. 1800.
PR/GE/110/3/23.
PR/GE/110/5/18: 18 Jun. 1756.
PR/GE/110/11/2: 1761.
PR/GE/110/11/25: 13 Apr. 1761.
PR/GE/112/13/5–7.
PR/GE/118/8/693–695.
PR/GE/118/11/935.
PR/GE/118/11/937–948.

Shropshire Archives

MS 552/12/153: Styche Hall inventory (1825).
MS C20/2629/1: 'A list of patents granted under the old law 1617 to 1852'.

Trinity College Dublin

MS 3951: Conolly Papers.
MS 10528: 'Ware household accounts book, 1740–86'.

The Ulster Museum

Templeton MSS, S54 (1806).

University of Nottingham Special Collections

PWE5.

Warwickshire County Record Office

Mordaunt Family of Walton Papers, CR1368/vol. 1.

Print

The annual biography and obituary for the year 1818, vol. 2 (London, 1818).

Anon., 'An historical account of Irish almanacks', *Irish magazine, or monthly asylum for neglected biography* (January 1810).

Boissier de Sauvages, Abbé, Joseph Crukshank, Isaac Collins, Jonathan Odell, Asa M. Stackhouse and Samuel Pullein, *Directions for the breeding and management of silk-worms: Extracted from the treatise of the Abbé Boissier de Sauvages, and Pullein* (Philadelphia, PA: Joseph Crukshank and Isaac Collins, 1770).

Bolton, James, *Harmonia ruralis; Or, an essay towards a natural history of British song birds*, 2 vols. (Halifax, 1794–6).

Burke's landed gentry, vol. 2 (London, 1847).

Chauvin, Étienne, *Lexicon philosophicum*, vol. 2 (Leeuwarden, 1713; facsimile repr. Düsseldorf: Stern-Verlag Janssen, 1967).

Dayes, Edward, 'Essay on the usefulness of drawing', *Belfast news-letter* (19 Jan. 1809).

The Dublin Society's weekly observations, vol. 1, no. 1: 4 Jan. 1736–37 (Dublin, 1739).

Gentleman's Magazine, vol. 10 (Feb. 1740); vol. 10 (Mar. 1740); vol. 10 (Jun. 1740); vol. 10 (Sep. 1740); vol. 10 (Oct. 1740); vol. 12 (Feb.

1742); vol. 12 (Apr. 1742); vol. 12 (May 1742); vol. 24 (Feb. 1754); vol. 24 (Mar. 1754); vol. 24 (Apr. 1754); vol. 26 (Apr. 1756); vol. 51 (1781); vol. 52 (Feb. 1782); vol. 68 (May 1790); vol. 62, pt. 2 (1791); vol. 64, pt. 2 (Jul. 1794); vol. 78, pt. 2 (Dec. 1808); vol. 87, pt. 1 (Apr. 1817).

Glauber, John Rudolph, *The works of the highly experienced and famous chymist, John Rudolph Glauber: Containing, great variety of choice secrets in medicine and alchymy* (London, 1689).

Harris, James, *Philosophical arrangements* (London, 1775).

Makin, Bathsua, *An essay to revive the ancient education of gentlewomen* (London, 1673).

Proceedings of the Dublin Society, vols 1–2 (15 Mar. 1764–2 Oct. 1766); vol. 5 (Oct. 1768–Jul. 1769); vol. 6 (Oct. 1769–Aug. 1770); vol. 10 (Oct. 1773–Aug. 1774); vol. 12 (Nov. 1775–Jun. 1776); vol. 37 (6 Nov. 1800–30 Jul. 1801); vol. 38 (5 Nov. 1801–26 Aug. 1802); vol. 39 (4 Nov. 1802–11 Aug. 1803); vol. 41 (1 Nov. 1804–15 Aug. 1805).

Pullein, Samuel, *Some hints intended to promote the culture of silkworms in Ireland* (Dublin, 1750).

Rhodes, Henrietta, *Rosalie: Or the castle of Montalabretti* (Richmond, 1811).

Rhodes, Henrietta, *Poems and miscellaneous essays* (Brentford, 1814).

Royal Society, *Philosophical Transactions*, vol. 46 (1749–50); vol. 51 (1759–60).

Caitlin, James and James Short, 'An account of the eclipse of the moon, June 8 1750', *Philosophical Transactions*, 46:496 (1749–50), pp. 523–5.

Skinner and Company, *A catalogue of the Portland Museum* (London, 1786).

Townley, Richard, *A journal kept in the Isle of Man*, vols. I–II (Whitehaven, 1791).

Transactions of the American Philosophical Society, vol. 1 (1 Jan. 1769–1 Jan. 1771); vol. 2 (1786).

Bartram, M., 'Observations on the native silk worms of north-America, by Mr. Moses Bartram. Read before the Society, March 11, I 768', *Transactions of the American Philosophical Society*, vol. 1 (1 Jan. 1769–1 Jan. 1771), pp. 224–30.

Gilpin, J., 'Observations on the annual passage of herrings', *Transactions of the American Philosophical Society*, vol. 2 (1786), pp. 236–9.

Nicola, Lewis, 'An easy method of preserving subjects in spirits', *Transactions of the American Philosophical Society*, vol. 1 (1769–71), pp. 244–6.

Transactions of the Dublin Society, vol. 1, pt. 2 (1799); vol. 2, pt. 1 (1800); vol. 2, pt. 2 (1801).

Transactions of the Society, instituted at London, for the Encouragement of Arts, Manufactures, and Commerce: vol. 2 (1784); vol. 4 (1786); vol. 5 (1787); vol. 11 (1793); vol. 14 (1796).

Marten, J., 'Papers in colonies and trade', *Transactions of the Society, instituted at London, for the Encouragement of Arts, Manufactures and Commerce*, vol. 11 (1793).

Williams, Ann, *Original poems and imitations* (London, 1773).

Online

Old Bailey proceedings online (www.oldbaileyonline.org, version 8.0, 11 November 2021), January 1829, trial of Benjamin Williams, John Brinkley (t18290115-47).

Old Bailey proceedings online (www.oldbaileyonline.org, version 8.0, 11 November 2021), October 1784, trial of Robert Artz, Thomas Gore (t17841020-9).

Secondary sources

Adamson, Glenn, *Craft: An American history* (London: Bloomsbury Publishing, 2021).

Alcock, N. W., *People at home: Living in a Warwickshire village, 1500–1800* (Chichester: Phillimore, 1993).

Allen, David, *Commonplace books and reading in Georgian England* (Cambridge: Cambridge University Press, 2010).

Allen, Katherine, 'Hobby and craft: Distilling household medicine in eighteenth-century England', *Early Modern Women: An Interdisciplinary Journal*, 11:1 (2016), pp. 90–114.

Anderson, R. G. W., M. L. Caygill, A. G. MacGregor and L. Syson (eds), *Enlightening the British: Knowledge, discovery and the museum in the eighteenth century* (London: British Museum Press, 2003).

Anishanslin, Zara, 'Unravelling the Silk Society's directions for the breeding and management of silk-worms', *Commonplace: The Journal of American Life*, 14:1 (2013): http://commonplace.online/article/unraveling-silk-society/ (accessed 19 August 2021).

Arizpe, Evelyn and Morag Styles with Shirley Brice Heath, *Reading lessons from the eighteenth century: Mothers, children and texts* (Shenstone: Pied Piper, 2006).

Aynsley, Jeremy and Charlotte Grant (eds), *Imagined interiors: Representing the domestic interior since the Renaissance* (London: V&A Publications, 2006).

Baigent, Elizabeth, 'Bullock, William (bap. 1773, d. 1849) naturalist and antiquary', *Oxford dictionary of national biography online* (Oxford: Oxford University Press, 2004): https://doi.org/10.1093/ref:odnb/3923

Barker, Hannah and Elaine Chalus (eds), *Gender in eighteenth-century England: Roles, representations, and responsibilities* (New York: Longman, 1997).

Barnard, Toby, *Brought to book: Print in Ireland, 1680–1784* (Dublin: Four Courts Press, 2017).

Barstow, Carol, *In Grandmother Gell's kitchen: A selection of recipes used in the eighteenth century* (Nottingham: Nottingham County Council, 2009).

Baudino, Isabelle, Jacques Carré and Cécile Révauger (eds), *The invisible woman: Aspects of women's work in eighteenth-century Britain* (Aldershot: Ashgate, 2005).

Bellanca, Mary E., 'Science, animal sympathy, and Anna Barbauld's "The mouse's petition"', *Eighteenth-Century Studies*, 37:1 (2003), pp. 47–67.

Bellanca, Mary E., *Daybooks of discovery: Nature diaries in Britain, 1770–1870* (London: University of Virginia Press, 2007).

Benedict, Barbara M., *Curiosity: A cultural history of early modern inquiry* (Chicago, IL: University of Chicago Press, 2001).

Benedict, Barbara M., 'Collecting trouble: Sir Hans Sloane's literary reputation in eighteenth-century Britain', *Eighteenth-Century Life*, 36:2 (2012), pp. 111–42.

Bennett, Andrew J., 'Expressivity: The Romantic theory of authorship' in Patricia Waugh (ed.), *Literary theory and criticism: An Oxford guide* (Oxford: Oxford University Press, 2006), pp. 48–58.

Benson, Charles, 'The Irish trade' in Michael F. Suarez and Michael L. Turner (eds), *The Cambridge history of the book in Britain*, vol. 5, 1695–1830 (Cambridge: Cambridge University Press, 2009), pp. 366–82.

Berg, Maxine, *Luxury & pleasure in eighteenth-century Britain* (Oxford: Oxford University Press, 2005).

Berry, Helen, *Gender, society and print culture in late Stuart England: The cultural world of the* Athenian Mercury (Aldershot: Ashgate, 2003).

Bertucci, Paolo, 'Sparks in the dark: The attraction of electricity in the eighteenth century', *Endeavour*, 31:3 (2007), pp. 88–93.

Blair, Ann, *The Theater of Nature: Jean Bodin and Renaissance science* (Princeton, NJ: Princeton University Press, 1997).

Blunt, Alison and Eleanor John, 'Domestic practice in the past: Historical sources and methods', *Home Cultures*, 11:3 (2014), pp. 269–74.

Botonaki, Effie, 'Seventeenth-century English women's spiritual diaries: Self-examination, covenanting, and account keeping', *The Sixteenth Century Journal*, 30:1 (1999), pp. 3–21.

Bourdieu, Pierre, *Outline of a theory of practice*, trans. Richard Nice (Cambridge: Cambridge University Press, 1977).
Bowen, H. V., 'Clive, Margaret, Lady Clive of Plassey (1735–1817)', *Oxford dictionary of national biography online* (Oxford: Oxford University Press, 2004): https://doi.org/10.1093/ref:odnb/63502
Bray, Francesca, *Technology and gender: Fabrics of power in late Imperial China* (Berkeley, CA: University of California Press, 1997).
Brears, Peter, 'Behind the green baize door' in Pamela A. Sambrook and Peter Brears (eds), *The country house kitchen, 1650–1900* (Stroud: The History Press, 2010), pp. 30–76.
Brears, Peter, 'The ideal kitchen in 1864' in Pamela A. Sambrook and Peter Brears (eds), *The country house kitchen, 1650–1900* (Stroud: The History Press, 2010), pp. 11–29.
Brewer, John and Roy Porter (eds), *Consumption and the world of goods* (London: Routledge, 1994).
Brown, Michael, *The Irish Enlightenment* (Cambridge: Harvard University Press, 2016).
Bruchac, Margaret M., 'Indigenous knowledge and traditional knowledge' in Claire Smith (ed.), *Encyclopedia of global archaeology* (New York: Springer Science and Business Media, 2014), pp. 3814–24.
Brück, Mary T., *Women in early British and Irish astronomy: Stars and satellites* (London: Springer, 2009).
Capp, Bernard, *Astrology and the popular press: English almanacs, 1500–1800* (London: Faber, 1979).
Carroll, Berenice A., 'The politics of "originality": Women and the class system of the intellect', *Journal of Women's History*, 2:2 (1990), pp. 136–63.
Clark, Peter, *Sociability and urbanity: Clubs and societies in the eighteenth-century city* (Leicester: Victorian Studies Centre, 1986).
Clark, Timothy J., *The sight of death: An experiment in art writing* (New Haven, CT and London: Yale University Press, 2006).
Clarke, Norma, *The rise and fall of the woman of letters* (London: Pimlico, 2004).
Coen, Deborah R., 'The common world: Histories of science and intimacy', *Modern Intellectual History*, 11:2 (2014), pp. 417–38.
Cohen, Michèle, 'Gender and the public private debate on education in the long eighteenth century' in Richard Aldrich (ed.), *Public or private education? Lessons from history* (London: Routledge, 2004), pp. 15–35.
Cohen, Michèle, '"To think, to compare, to combine, to methodise": Notes towards rethinking girls' education in the eighteenth century' in Sarah Knott and Barbara Taylor (eds), *Women, gender and enlightenment* (Basingstoke: Palgrave Macmillan, 2005), pp. 224–42.

Cohen, Michèle, '"Familiar conversation": the role of the "familiar format" in education in eighteenth- and nineteenth-century England' in Jill Shefrin and Mary Hilton (eds), *Educating the child in enlightenment Britain: Beliefs, cultures, practices* (London: Routledge, 2009), pp. 99–117.

Cohen, Michèle, 'French conversation of "glittering gibberish"? Learning French in eighteenth-century England' in Natasha Glaisyer and Sara Pennell (eds), *Didactic literature in England, 1500–1800* (London: Routledge, 2017), pp. 99–117.

Colin, Finn, *Science studies as naturalized philosophy* (Dordrecht: Springer, 2011).

Cooper, Alix, 'Homes and households' in Katharine Park and Lorraine Daston (eds), *The Cambridge history of science*, vol. 3 (Cambridge: Cambridge University Press, 2006), pp. 224–37.

Cooper, Alix, 'Afterword' in Donald L. Opitz, Staffan Bergwik and Brigitte Van Tiggelen (eds), *Domesticity in the making of modern science* (London: Palgrave Macmillan, 2016), pp. 281–7.

Costa, Shelley, 'The "Ladies' diary": Gender, mathematics and civil society in early-eighteenth-century England', *Osiris*, 17 (2002), pp. 49–73.

Coulton, Richard, '"The darling of the Temple-Coffee-House Club": Science, sociability and satire in early eighteenth-century London', *Journal for Eighteenth-Century Studies*, 35:1 (2012), pp. 43–65.

Coulton, Richard, '"What he hath gather'd together shall not be lost": Remembering James Petiver', *Notes and Records*, 74 (2020), pp. 189–211.

Craciun, Adriana and Simon Schaffer, *The material cultures of enlightenment arts and sciences* (London: Palgrave Macmillan, 2016).

Creager, Angela N. H., Mathias Grote and Elaine Leong, 'Learning by the book: Manuals and handbooks in the history of science', *British Journal for the History of Science: Themes*, 5 (2020), pp. 1–13.

Cromley, Elizabeth Collins, *The food axis: Cooking, eating, and the architecture of American houses* (Charlottesville, VA: University of Virginia Press, 2010).

Crowley, John E., *The invention of comfort: Sensibilities and design in early modern Britain and early America* (Baltimore, MD: Johns Hopkins University Press, 2001).

Curth, Louise Hill, 'The medical content of English almanacs, 1640–1700', *Journal of the History of Medicine*, 60:3 (2005), pp. 255–82.

Dacome, Lucia, 'Noting the mind: Commonplace books and the pursuit of the self in eighteenth-century Britain', *Journal of the History of Ideas*, 65:4 (2004), pp. 603–25.

Danaher, Kevin, *Ireland's traditional houses* (Dublin: Bord Fáilte, 1993).

Daskalova, Krassimira, Mary O'Dowd and Daniela Koleva, 'Introduction', *Women's History Review*, Special issue: Gender and the cultural production of knowledge, 20:4 (2011), pp. 487–9.

Daston, Lorraine, 'The empire of observation 1600–1800' in Lorraine Daston and Elizabeth Lunbeck (eds), *Histories of scientific observation* (Chicago, IL: University of Chicago Press, 2011), pp. 81–113.

Daston, Lorraine, 'The history of science and the history of knowledge', *KNOW: A Journal on the Formation of Knowledge*, 1:1 (2017), pp. 131–54.

Daston, Lorraine and Peter Galison, *Objectivity* (New York: Zone Books, 2007).

Daston, Lorraine and Elizabeth Lunbeck (eds), *Histories of scientific observation* (Chicago, IL: Chicago University Press, 2011).

Daston, Lorraine and Katharine Park, *Wonders and the order of nature, 1150–1750* (New York: Zone Books, 1998).

Daunton, Martin J., *Progress and poverty: An economic and social history of Britain, 1700–1850* (Oxford: Oxford University Press, 1995).

Dear, Peter, 'The meanings of experience' in Katharine Park and Lorraine Daston (eds), *The Cambridge history of science*, vol. 3 (Cambridge: Cambridge University Press, 2006), pp. 106–31.

de Certeau, Michel, *The practice of everyday life*, trans. Steven F. Rendall (Berkeley, CA: University of California Press, 1988).

de la Bellacasa, Maria Puig, *Matters of care: Speculative ethics in more than human worlds* (Minneapolis, MN: University of Minnesota Press, 2017).

de Montluzin, Emily Lorraine, *Daily life in Georgian England as reported in the Gentleman's Magazine* (Lampeter: The Edwin Mellen Press, 2002).

de Saussure, Ferdinand, *Cours de linguistique générale* (Paris: Payot, 1916).

de Vries, Jan, *Consumer behavior and the household economy, 1650 to the present* (Cambridge: Cambridge University Press, 2008).

Delbourgo, James, *Collecting the world: The life and curiosity of Hans Sloane* (London: Penguin Books, 2017).

Delbourgo, J. and N. Dew (eds), *Science and empire in the Atlantic world* (London: Routledge, 2008).

Delbourgo, James, Kapil Raj, Lissa Roberts and Simon Schaffer (eds), *The brokered world: Go-betweens and global intelligence, 1770–1820* (Sagamore Beach, MA: Science History Publications, 2009).

Deleuze, Gilles, *Difference and repetition* (London: Bloomsbury Academic, 2014).

Dillon, Brian and Marina Warner, *Curiosity: Art and the pleasures of knowing* (London: Hayward Publishing, 2013).

DiMeo, Michelle, 'Lady Ranelagh's book of kitchen-physick? Reattributing authorship for Wellcome Library MS 1340', *Huntington Library Quarterly*, 77:3 (2014), pp. 331–46.

DiMeo, Michelle, '"Such a sister became such a brother": Lady Ranelagh's influence on Robert Boyle', *Intellectual History Review*, 25:1 (2015), pp. 21–36.

Drury, Paul J. (with a major contribution by Richard Simpson), *Hill Hall: A singular house devised by a Tudor intellectual* (London: Society of Antiquaries, 2009).

Dunlevy, Mairead, *Pomp and poverty: A history of silk in Ireland* (London: Yale University Press, 2011).

Dupré, Sven and Christoph Herbert Lüthy (eds), *Silent messengers: The circulation of material objects of knowledge in the early modern Low Countries* (Berlin: LIT Verlag, 2011).

Easterby-Smith, Sarah, 'Recalcitrant seeds: Material culture and the global history of science', *Past and Present*, supplement 14 (2019), pp. 215–42.

Eddy, M. D., 'Tools for reordering: Commonplacing and the space of words in Linnaeus's *Philosophia botanica*', *Intellectual History Review*, 20:2 (2010), pp. 227–52.

Edgerton, David, *The shock of the old: Technology and global history since 1900* (London: Profile, 2008).

Edmondson, John, 'New insights into John Bolton of Halifax', *Mycologist*, 9:4 (1995), pp. 174–8.

Ellis, Markman, *The coffee house: A cultural history* (London: Weidenfeld and Nicolson, 2011).

Engeström, Yrjö and David Middleton (eds), *Cognition and communication at work* (Cambridge: Cambridge University Press, 1996).

Evans, R. J. W. and Alexander Marr (eds), *Curiosity and wonder from the Renaissance to the Enlightenment* (Aldershot: Ashgate, 2006).

Fara, Patricia, *Sympathetic attractions: Magnetic practices, beliefs, and symbolism in eighteenth-century England* (Princeton, NJ: Princeton University Press, 1996).

Fara, Patricia, *Pandora's breeches: Women, science and power in the Enlightenment* (London: Pimlico, 2004).

Ferguson, Margaret W., *Dido's daughters: Literacy, gender, and empire in early modern England and France* (London: University of Chicago Press, 2003).

Findlen, Paula, 'Sites of anatomy, botany, and natural history' in Katharine Park and Lorraine Daston (eds), *The Cambridge history of science*, vol. 3 (Cambridge: Cambridge University Press, 2006), pp. 272–89.

Findlen, Paula (ed.), *Empires of knowledge: Scientific networks in the early modern world* (London: Routledge, 2018).

Fisher, James, 'The master should know more: Book-farming and the conflict over agricultural knowledge', *Cultural and Social History*, 15:3 (2018), pp. 315–31.

Fleming, David, 'Cycles, seasons and the everyday in mid-eighteenth-century provincial Ireland' in Raymond Gillespie and R. F. Foster (eds), *Irish provincial cultures in the long eighteenth century: Making the middle sort, essays for Toby Barnard* (Dublin: Four Courts Press, 2012), pp. 133–54.

Fox, Celina, *The arts of industry in the age of enlightenment* (New Haven, CT: Yale University Press, 2009).

Friedewald, Boris, *A butterfly journey: Maria Sibylla Merian artist and scientist* (Munich: Prestel, 2015).

Galison, Peter, 'Limits of localism: The scale of sight' in Wendy Doniger, Peter Galison and Susan Neiman (eds), *What reason promises: Essays on reason, nature and history* (Berlin: De Gruyter, 2016), pp. 155–70.

Gater, G. H. and F. R. Hiorns (eds), 'Hemmings Row and Castle Street' in *Survey of London: Vol. 20, St Martin-in-The-Fields, Pt III: Trafalgar Square and Neighbourhood* (London: British History Online, 1940), pp. 112–14: www.british-history.ac.uk/survey-london/vol20/pt3/pp112-114 (accessed 11 November 2021).

Gaukroger, Stephen, *Francis Bacon and the transformation of early-modern philosophy* (Cambridge: Cambridge University Press, 2001).

George, Sam, *Botany, sexuality and women's writing, 1760–1830: From modest shoot to forward plant* (Manchester: Manchester University Press, 2007).

Gibson, Susannah, *Animal, vegetable, mineral?: How eighteenth-century science disrupted the natural order* (Oxford: Oxford University Press, 2015).

Gillespie, Raymond, 'Climate, weather and social change in seventeenth-century Ireland', *Proceedings of the Royal Irish Academy: Archaeology, Culture, History, Literature*, 120C (2020), pp. 263–71.

Glaisyer, Natasha and Sara Pennell (eds), *Didactic literature in England, 1500–1800* (London: Routledge, 2017).

Golinski, Jan, *Science as public culture: Chemistry and enlightenment in Britain, 1760–1820* (Cambridge: Cambridge University Press, 1992).

Golinski, Jan, *British weather and the climate of enlightenment* (Chicago, IL: University of Chicago Press, 2007).

Goodbody, Olive C., *Guide to Irish Quaker records, 1654–1860* (Dublin: Stationery Office for the Irish Manuscripts Commission, 1967).

Goss, David A., 'Benjamin Martin (1704–1782) and his writings on the eye and eyeglasses', *Hindsight*, 41:2 (2010), pp. 41–8.

Guerrini, Anita, 'The ghastly kitchen', *History of Science*, 54:1 (2016), pp. 71–97.

Hamlett, Jane, 'The British domestic interior and social and cultural history', *Cultural and Social History*, 6:1 (2009), pp. 97–107.

Handley, Sasha, 'Objects, emotions and an early modern bed-sheet', *History Workshop Journal*, 85 (2018), pp. 169–94.

Hannan, Leonie, 'Collaborative scholarship on the margins: An epistolary network', *Women's Writing*, 21:3 (2014), pp. 290–315.

Hannan, Leonie, *Women of letters: Gender, writing and the life of the mind in early modern England* (Manchester: Manchester University Press, 2016).

Hardyment, Christina, *Home comfort: A history of domestic arrangements* (London: Viking Penguin in association with the National Trust, 1992).

Hardyment, Christina (ed.), *The housekeeping book of Susanna Whatman* (London: The National Trust, 1992).

Harkness, Deborah E., 'Managing an experimental household: The Dees of Mortlake and the practice of natural philosophy', *Isis*, 88:2 (1997), pp. 247–62.

Harris, Bob, 'Print culture' in H. T. Dickenson (ed.), *A companion to eighteenth-century Britain* (Oxford: Blackwell Publishing, 2002), pp. 283–93.

Harris, Frances and Michael Hunter (eds), *John Evelyn and his milieu* (London: The British Library, 2003).

Harris, John R., 'Skills, coal and British industry in the eighteenth century', *History*, 61 (1976), pp. 167–82.

Harrison, Richard S., *Dr John Rutty (1698–1775) of Dublin: A Quaker polymath in the Enlightenment* (Dublin: Original Writing, 2011).

Harvey, Karen, *The little Republic: Masculinity and domestic authority in eighteenth-century Britain* (Oxford: Oxford University Press, 2012).

Harvey, Karen, 'Oeconomy and the eighteenth-century house: A cultural history of social practice', *Home Cultures*, 11:3 (2014), pp. 375–90.

Havard, Lucy J., '"Preserve or perish": Food preservation practices in the early modern kitchen', *Notes and Records*, 74 (2020), pp. 5–33.

Haynes, Clare, 'A "natural" exhibitioner: Sir Ashton Lever and his *Holosphsikon*', *British Journal for Eighteenth-Century Studies*, 24 (2001), pp. 1–14.

Hayton, David, 'Review: Marie-Louise Legg (ed.), *The Synge letters: Bishop Edward Synge to his daughter Alicia: Roscommon to Dublin, 1746–1752* (Dublin: Lilliput Press, 1996)', *Irish Historical Studies*, 30:119 (1997), pp. 479–80.

Hecht, Joseph J., *The domestic servant class in eighteenth-century England* (London: Routledge & Kegan Paul, 1956).

Hellawell, Philippa, '"The best and most practical philosophers": Seamen and the authority of experience in early modern science', *History of Science*, 58:1 (2019), pp. 1–23.

Hendriksen, Marieke, '"Art and technique always balance the scale": German philosophies of sensory perception, taste, and art criticism, and the rise of the term Technik, ca. 1735–ca. 1835', *History of Humanities*, 2:1 (2017), pp. 201–19.

Hendriksen, Marieke, 'Review of *The structures of practical knowledge*, edited by Matteo Valleriani', *Ambix*, 66:1 (2019), pp. 88–90.

Henry, John, *Knowledge is power: Francis Bacon and the method of science* (Cambridge: Icon Books, 2002).

Herbert, Amanda E., *Female alliances: Gender, identity, and friendship in early modern Britain* (New Haven, CT: Yale University Press, 2014).

Herries Davies, Gordon L., 'The Physico-Historical Society of Ireland, 1744–1752', *Irish Geography*, xii (1979), pp. 92–8.

Hickman, Clare, 'The garden as a laboratory: The role of domestic gardens as places of scientific exploration in the long 18th century', *Post-Medieval Archaeology*, 48:1 (2014), pp. 229–47.

Hickman, Clare, *The doctor's garden: Medicine, science, and horticulture in Britain* (New Haven, CT: Yale University Press, 2022).

Hill, Bridget, *Women, work and sexual politics in eighteenth-century England* (Oxford: Basil Blackwell, 1989).

Hill, Bridget, *Servants: English domestics in the eighteenth century* (Oxford: Clarendon, 1996).

Hodacs, Hanna, Kenneth Nyberg and Stéphanie van Damme (eds), *Linnaeus, natural history and the circulation of knowledge* (Oxford: Voltaire Foundation, 2018).

Horn, Jeff, Leonard N. Rosenband and Merritt Roe Smith, *Reconceptualizing the industrial revolution* (Cambridge, MA: MIT Press, 2010).

Howes, Anton, *Arts and minds: How the Royal Society of Arts changed a nation* (Oxford: Princeton University Press, 2020).

Hubbard, Eleanor, 'Reading, writing, and initialing: Female literacy in early modern London', *Journal of British Studies*, 54:3 (2015), pp. 553–77.

Humphries, Jane, 'The wages of women in England, 1260–1850', *The Journal of Economic History*, 75:2 (2015), pp. 405–47.

Humphries, Jane and Jacob Weisdorf, 'Unreal wages? Real income and economic growth in England, 1260–1850', *The Economic Journal*, 129:623 (2019), pp. 2867–87.

Hunt, Margaret, *The middling sort: Commerce, gender, and the family in England, 1680–1780* (London: University of California Press, 1996).

Hunter, Lynette and Sarah Hutton (eds), *Women, science and medicine, 1500–1700: Mothers and sisters of the Royal Society* (Stroud: Sutton Publishing Limited, 1997).

Hutchins, Edwin, *Cognition in the wild* (Cambridge, MA: MIT Press, 1995).

Iliffe, Robert, 'Material doubts: Hooke, artisan culture and the exchange of information in 1670s London', *British Journal for the History of Science*, 28 (1995), pp. 285–318.

Ingold, Tim, *The perception of the environment: Essays in livelihood, dwelling and skill* (London: Routledge, 2000).

Jones, Peter M., *Agricultural enlightenment: Knowledge, technology, and nature, 1750–1840* (Oxford: Oxford University Press, 2016).

Jordanova, Ludmilla, 'Gender and the historiography of science', *The British Journal of the History of Science*, 26:4 (1993), pp. 469–83.

Kelly, James, 'Climate, weather and society in Ireland in the long eighteenth-century: The experience of the later stages of the Little Ice Age', *Proceedings of the Royal Irish Academy: Archaeology, Culture, History, Literature*, 120C (2020), pp. 273–324.

Kelly, James and Martyn J. Powell (eds), *Clubs and societies in eighteenth-century Ireland* (Dublin: Four Courts Press, 2010).

Kennedy, Neil, *Curiosity in early modern Europe: World histories* (Wiesbaden: Harrassowitz, 1998).

Klein, Randolph Shipley, 'Moses Bartram (1732–1809)', *Quaker History*, 57:1 (1968), pp. 28–34.

Klein, Ursula, *Experiments, models, paper tools: Cultures of organic chemistry in the nineteenth century* (Stanford, CA: Stanford University Press, 2003).

Klein, Ursula, 'The laboratory challenge: Some revisions of the standard view of early modern experimentation', *Isis*, 99:4 (2008), pp. 769–82.

Klein, Ursula and Wolfgang Lefèvre, *Materials in eighteenth-century science: A historical ontology* (London: MIT, 2007).

Knott, Sarah and Barbara Taylor (eds), *Women, gender and enlightenment* (Basingstoke: Palgrave Macmillan, 2005).

Kripke, Saul, *Naming and necessity* (Cambridge, MA: Harvard University Press, 1980).

Lake, Crystal B., *Artifacts: How we think and write about found objects* (Baltimore, MD: Johns Hopkins University Press, 2020).

Landry, Donna, 'Green languages? Women poets as naturalists in 1653 and 1807', *Huntington Library Quarterly*, 63:4 (2000), pp. 467–89.

Latour, Bruno, *Science in action: How to follow scientists and engineers through society* (Cambridge, MA: Harvard University Press, 1987).

Latour, Bruno, *Pandora's hope: Essays on the reality of science studies* (Cambridge, MA: Harvard University Press, 1999).

Latour, Bruno, *Reassembling the social: An introduction to actor-network-theory* (Oxford: Oxford University Press, 2005).

Lee, Grace Lawless, *The Huguenot settlements in Ireland* (Berwyn Heights, MD: Heritage Books, 2008).

Legg, Marie-Louise (ed.), *The Synge letters: Bishop Edward Synge to his daughter Alicia, Roscommon to Dublin 1746–1752* (Dublin: Lilliput Press, 1996).

Leighton, Ann, *American gardens in the eighteenth century: 'For use or for delight'* (Boston, MA: Houghton Mifflin, 1986).

Lemire, Beverly, *The business of everyday life: Gender, practice and social politics in England, c. 1600–1900* (Manchester: Manchester University Press, 2012).

Leong, Elaine, 'Collecting knowledge for the family: Recipes, gender and practical knowledge in the early modern English household', *Centaurus*, 55:2 (2013), pp. 81–103.

Leong, Elaine, *Recipes and everyday knowledge: Medicine, science and the household in early modern England* (Chicago, IL: University of Chicago Press, 2018).

Leong, Elaine and Alisha Rankin (eds), *Secrets and knowledge in medicine and science, 1500–1800* (Farnham: Ashgate, 2011).

Leong, Elaine and Alisha Rankin, 'Testing drugs and trying cures: Experiment and medicine in medieval and early modern Europe', *Bulletin of the History of Medicine*, 91:2 (2017), pp. 157–82.

Lester, Alan, 'Spatial concepts and the historical geographies of British colonialism' in Andrew S. Thompson (ed.), *Writing imperial histories* (Manchester: Manchester University Press, 2013), pp. 118–42.

Lightman, Bernard, 'Marketing knowledge for the general reader: Victorian popularizers of science', *Endeavour*, 24:3 (2000), pp. 100–6.

Livesey, James, *Civil society and empire: Ireland and Scotland in the eighteenth-century Atlantic world* (London: Yale University Press, 2009).

Livingstone, David, *Putting science in its place: Geographies of scientific knowledge* (Chicago, IL: University of Chicago Press, 2003).

Long, Pamela O., *Artisan/practitioners and the rise of the new sciences, 1400–1600* (Corvallis, OR: Oregon State University Press, 2014).

Mack, Robert L., *The genius of parody: Imitation and originality in seventeenth- and eighteenth-century English literature* (Basingstoke: Palgrave Macmillan, 2007).

Magennis, Eoin, '"Land of milk and honey": The Physico-Historical Society, improvement and the surveys of mid-eighteenth-century Ireland', *Proceedings of the Royal Irish Academy: Archaeology, Culture, History, Literature*, 102C:6 (2002), pp. 199–217.

Marples, Alice, 'Medical practitioners as collectors and communicators of natural history in Ireland, 1680–1750' in John Cunningham (ed.), *Early modern Ireland and the world of medicine: Practitioners, collectors and contexts* (Manchester: Manchester University Press, 2019), pp. 147–64.

Marples, Alice, 'James Petiver's "joynt-stock": Middling agency in urban collecting networks', *Notes and Records: The Royal Society Journal of the History of Science*, 74:2 (2020), pp. 239–58.

Marshall, Dorothy, *The English domestic servant in history* (London: Historical Association, 1949).

Martin, Ann Smart, *Buying into the world of goods: Early consumers in backcountry Virginia* (Baltimore, MD: Johns Hopkins Press, 2008).

Mateus, Carla, 'Searching for historical meteorological observations on the island of Ireland', *Weather*, 76:5 (2021), pp. 160–5.

Mathias, Peter, 'Agriculture and the brewing and distilling industries in the eighteenth century', *The Economic History Review*, 5:2 (1952), pp. 249–57.

Mathias, Peter, *The brewing industry in England, 1700–1830* (Cambridge: Cambridge University Press, 1959).

McBride, Ian, 'The edge of enlightenment: Ireland and Scotland in the eighteenth century', *Modern Intellectual History*, 10:1 (2013), pp. 135–51.

McCann, Alison, 'A private laboratory at Petworth House, Sussex, in the late eighteenth century', *Annals of Science*, 40:6 (1983), pp. 635–55.

McKeon, Michael, *The secret history of domesticity: Public, private, and the division of knowledge* (Baltimore, MD: Johns Hopkins University Press, 2005).

McMenamin, Deirdre and Dougal Sheridan, 'Interpreting vernacular space in Ireland: A new sensibility', *Landscape Research*, 44:7 (2019), pp. 787–803.

Meacham, Sarah Hand, *Every home a distillery: Alcohol, gender, and technology in the colonial Chesapeake* (Baltimore, MD: Johns Hopkins Press, 2009).

Medeiros, Maria Franco Trindade, *Historical ethnobiology* (Amsterdam: Academic Press, 2020).

Meenan, James and Desmond Clarke (eds), *The Royal Dublin Society, 1731–1981* (Dublin: Gill and Macmillan Ltd, 1981).

Millman, P. M., 'Meteor news – telescopic meteor observations; the Ierofeevka meteorite; meteor heights determined in the XVIII century', *Journal of the Royal Astronomical Society of Canada*, 31 (1937), pp. 363–6.

Mistry, J., 'Indigenous knowledges' in Rob Kitchin and Nigel Thrift (eds), *International Encyclopaedia of Human Geography* (London: Elsevier Science, 2009), pp. 371–6.

Mitchell, Mark T., *Michael Polanyi: The art of knowing* (Wilmington, DE: ISI Books, 2006).

Mokyr, Joel, *The gifts of Athena: Historical origins of the knowledge economy* (Princeton, NJ: Princeton University Press, 2002).

Moss, Ann, *Printed commonplace-books and the structuring of Renaissance thought* (Oxford: Oxford University Press, 1996).

Mukerji, Chandra, 'Women engineers and the culture of the Pyrenees: Indigenous knowledge and engineering in seventeenth-century France' in Pamela H. Smith and Benjamin Schmidt (eds), *Making knowledge in early modern Europe: Practices, objects, and texts, 1400–1800* (Chicago, IL: Chicago University Press, 2007), pp. 19–44.

Mukerji, Chandra, 'The cultural power of tacit knowledge: Inarticulacy and Bourdieu's habitus', *American Journal of Cultural Sociology*, 2:3 (2014), p. 371 (pp. 348–75).

Muldrew, Craig, *Food, energy and the creation of industriousness: Work and material culture in agrarian England, 1550–1780* (Cambridge: Cambridge University Press, 2011).

Nevin, Monica, 'A County Kilkenny, Georgian household notebook', *The Journal of the Royal Society of Antiquaries of Ireland*, 109 (1979), pp. 5–19.

O'Dowd, Mary, *A history of women in Ireland, 1500–1800* (Harlow: Longman, 2005).

O'Reilly, Barry, 'Hearth and home: The vernacular house in Ireland from *c.* 1800', *Proceedings of the Royal Irish Academy: Archaeology, Culture, History, Literature*, 111C (2011), pp. 193–215.

Opitz, Donald L., Staffan Bergwik and Brigitte Van Tiggelen (eds), *Domesticity in the making of modern science* (London: Palgrave Macmillan, 2016).

Orlin, Lena Cowen, *Locating privacy in Tudor London* (Oxford: Oxford University Press, 2007).

Otter, Chris, 'Locating matter: the place of materiality in urban history' in Tony Bennett and Patrick Joyce (eds), *Material powers: Cultural studies, history and the material turn* (London: Routledge, 2010), pp. 38–59.

Otto Sibum, H., 'Science and the knowing body: Making sense of embodied knowledge in scientific experiment' in Sven Dupré, Anna Harris, Julia Kursell, Patricia Lulof and Maartje Stols-Wilcox (eds), *Reconstruction, replication and re-enactment in the humanities and social sciences* (Amsterdam: Amsterdam University Press, 2020), pp. 275–93.

Pal, Carol, *Republic of women: Rethinking the republic of letters in the seventeenth century* (Cambridge: Cambridge University Press, 2012).

Parrish, Susan Scott, *American curiosity: Cultures of natural history in the colonial British Atlantic world* (Chapel Hill, NC: University of North Carolina Press, 2006).

Pearce, Michael, 'Approaches to household inventories and household furnishing, 1500–1650', *Architectural Heritage*, 26 (2015), pp. 73–86.
Pearson, Jacqueline, *Women's reading in Britain, 1750–1835: A dangerous recreation* (Cambridge: Cambridge University Press, 1999).
Peck, Linda Levy, *Consuming splendor: Society and culture in seventeenth-century England* (Cambridge: Cambridge University Press, 2005).
Peck, Robert McCracken, 'Preserving nature for study and display' in Sue Ann Prince (ed.), *Stuffing birds, pressing plants, shaping knowledge: natural history in north America, 1730–1860* (Philadelphia, PA: American Philosophical Society, 2003), pp. 11–25.
Pennell, Sara, *The birth of the English kitchen, 1600–1850* (London: Bloomsbury Academic, 2016).
Pérez-Ramos, Antonio, *Francis Bacon's idea of science and the maker's knowledge tradition* (Oxford: Clarendon Press, 1988).
Pesic, Peter, 'Wrestling with Proteus: Francis Bacon and the "torture" of nature', *Isis*, 90 (1999), pp. 81–94.
Phillips, Nicola, *Women in business, 1700–1850* (Woodbridge: Boydell and Brewer, 2006).
Phillips, Patricia, *The scientific lady: A social history of women's scientific interests, 1520–1918* (London: Weidenfeld and Nicolson, 1990).
Plumwood, Val, *Feminism and the mastery of nature* (London: Routledge, 1993).
Polanyi, Michael, *Personal knowledge: Towards a post-critical philosophy* (New York, NY: Harper & Row, 1964).
Polanyi, Michael, *The tacit dimension* (Chicago, IL: Chicago University Press, 2009 [1966]).
Pollard, Mary, *A dictionary of members of the Dublin book trade, 1550–1800* (London: Bibliographical Society, 2000).
Pomian, Krzysztof, *Collectors and curiosities: Paris and Venice, 1500–1800*, trans. Elizabeth Wiles-Portier (Cambridge: Polity, 1990).
Poole, Robert, '"Give us our eleven days!": Calendar reform in eighteenth-century England', *Past & Present*, 149 (1995), pp. 95–139.
Poovey, Mary, *A history of the modern fact: Problems of knowledge in the sciences of wealth and society* (London: University of Chicago Press, 1998).
Porter, Roy, 'English society in the eighteenth century revisited' in Jeremy Black (ed.), *British politics and society from Walpole to Pitt 1742–89* (Basingstoke: Macmillan, 1990), pp. 29–52.
Potter, Jennifer, *Strange blooms: The curious lives and adventures of the John Tradescants* (London: Atlantic Books, 2014).
Prendergast, Amy, *Literary salons across Britain and Ireland in the long eighteenth century* (London: Palgrave Macmillan, 2015).

Priestley, Ursula and Penelope Corfield, 'Rooms and room use in Norwich housing, 1580–1730', *Post-Medieval Archaeology*, 16 (1982), pp. 93–123.

Probyn, C. T., *The sociable humanist: The life and works of James Harris, 1709–1780* (Oxford: Clarendon Press, 1991).

Raftery, Deirdre, *Women and learning in English writing, 1600–1900* (Dublin: Four Courts, 1997).

Randolph, John, *The house in the garden: The Bakunin family and the romance of Russian idealism* (London: Cornell University Press, 2007).

Rankin, Alisha, *Panaceia's daughters: Noblewomen and healers in early modern Germany* (Chicago, IL: University of Chicago Press, 2013).

Raven, James, Helen Small and Naomi Tadmor (eds), *The practice and representation of reading in England* (Cambridge: Cambridge University Press, 2007).

Ray, Meredith K., *Daughters of alchemy: Women and scientific culture in early modern Italy* (Cambridge, MA: Harvard University Press, 2015).

Reckwitz, Andreas, 'Toward a theory of social practices: A development in culturalist theorizing', *European Journal of Social Theory*, 5:2 (2002), pp. 243–63.

Roberts, Lissa, 'Practicing oeconomy during the second half of the long eighteenth century: An introduction', *History and Technology*, 30 (2014), pp. 133–48.

Roberts, Lissa L. and Simon Werrett (eds), *Compound histories: Materials, governance, and production, 1760–1840* (Leiden: Brill, 2018).

Roberts, Lissa L. and Simon Werrett, 'Introduction: "A more intimate acquaintance"' in Lissa L. Roberts and Simon Werrett (eds), *Compound histories: Materials, governance and production, 1760–1840* (Leiden: Brill, 2018), pp. 1–32.

Roberts, Lissa, Simon Schaffer and Peter Dear (eds), *The mindful hand: Inquiry and invention from the late Renaissance to early industrialisation* (Amsterdam: Edita KNAW, 2007).

Rothschild, Emma, *The inner life of empires: An eighteenth-century history* (Princeton, NJ: Princeton University Press, 2011).

Rule, John, *Albion's people: English society, 1714–1815* (London: Longman, 1992).

Salmon, Michael A., Peter Marren and Basil Harley, *The Aurelian legacy: British butterflies and their collectors* (Berkeley and Los Angeles, CA: University of California Press, 2000).

Sambrook, Pamela A. and Peter Brears (eds), *The country house kitchen, 1650–1900* (Stroud: The History Press, 2010).

Sanderson, Michael G., 'Daily weather in Dublin 1716–1734: The diary of Isaac Butler', *Weather*, 73:6 (2018), pp. 179–82.

Sayle, Charles E. (ed.), *A catalogue of the Bradshaw Collection of Irish books in the University Library Cambridge, vol. 1, books printed in Dublin by known printers, 1602–1882* (Cambridge: Cambridge University Press, 2014).

Schiebinger, Londa, 'Gender and natural history' in N. Jardine, J. A. Secord and E. C. Spary (eds), *Cultures of natural history* (Cambridge: Cambridge University Press, 1996), pp. 163–77.

Schiebinger, Londa, *Secret cures of slaves: People, plants, and medicine in the eighteenth-century Atlantic world* (Stanford, CA: Stanford University Press, 2017).

Schiffer, Margaret B., *Chester County, Pennsylvania inventories, 1684–1850* (Exton, PA: Schiffer Publishing, Ltd., 1974).

Schteir, Ann B., *Cultivating women, cultivating science: Flora's daughters and botany in England, 1760 to 1860* (London: Johns Hopkins Press Ltd., 1996).

Scott, Joan Wallach, 'Gender: a useful category of historical analysis', *The American Historical Review*, 91:5 (1986), pp. 1053–75.

Scott-Warren, Jason, 'Early modern bookkeeping and life-writing revisited: Accounting for Richard Stonley', *Past & Present*, 230:11 (2016), pp. 151–70.

Seaward, M., 'Bolton, James (*bap.* 1735, *d.* 1799)', *Oxford dictionary of national biography online* (Oxford: Oxford University Press, 2004): https://doi.org/10.1093/ref:odnb/2803

Secord, Anne, 'Science in the pub: Artisan botanists in early nineteenth-century Lancashire', *History of Science*, 32 (1994), pp. 269–315.

Secord, Anne, 'Elizabeth Gaskell and the artisan naturalists of Manchester', *The Gaskell Society Journal*, 19 (2005), pp. 34–51.

Secord, Anne, 'Coming to attention: a commonwealth of observers during the Napoleonic Wars' in Lorraine Daston and Elizabeth Lunbeck (eds), *Histories of scientific observation* (Chicago, IL: University of Chicago Press, 2011), pp. 421–44.

Secord, Anne, 'Introduction' in G. White (A. Secord (ed.)), *The natural history of Selbourne* (Oxford: Oxford University Press, 2013).

Secord, James A., 'Knowledge in transit', *Isis*, 95:4 (2004), pp. 654–72.

Seidel, Linda, 'Visual representation as instructional text: Jan van Eyck and the Ghent altarpiece' in Pamela H. Smith and Benjamin Schmidt, *Making knowledge in early modern Europe: Practices, objects, and texts, 1400–1800* (Chicago, IL: Chicago University Press, 2007), pp. 45–67.

Sennett, Richard, *The craftsman* (London: Allen Lane, 2008).

Shanahan, Madeline, '"Whipt with a twig rod": Irish manuscript recipe books as sources for the study of culinary material culture, c. 1660 to 1830', *Proceedings of the Royal Irish Academy: Section C: Archaeology,*

Celtic Studies, History, Linguistics, Literature, 115C (2015), pp. 197–218.

Shapin, Steven, 'The house of experiment in seventeenth-century England', *Isis*, 79:3 (1988), pp. 373–404.

Shapin, Steven, *A social history of truth: Civility and science in seventeenth-century England* (London: University of Chicago Press, 1994).

Shapin, Steven, *Never pure: Historical studies of science as if it was produced by people with bodies, situated in time, space, culture, and society and struggling for credibility and authority* (Baltimore, MD: Johns Hopkins University Press, 2010).

Sheehan, Jonathan and Dror Wahrman, *Invisible hands: Self-organisation and the eighteenth century* (Chicago, IL: University of Chicago Press, 2015).

Silver, Sean, *The mind is a collection: Case studies in eighteenth-century thought* (Philadelphia, PA: University of Pennsylvania Press, 2015).

Simmons, R. C., 'ABCs, almanacs, ballads, chapbooks, popular piety and textbooks' in John Barnard and D. F. McKenzie (eds), *The Cambridge history of the book in Britain*, vol. 4 (Cambridge: Cambridge University Press, 2014), pp. 504–13.

Sloboda, Stacey, 'Displaying materials: Porcelain and natural history in the Duchess of Portland's museum', *Eighteenth-Century Studies*, 43:4 (2010), pp. 455–72.

Smith, Pamela H., *The body of the artisan: Art and experience in the scientific revolution* (Chicago, IL: University of Chicago Press, 2004).

Smith, Pamela H., 'Vermilion, mercury, blood, and lizards: matter and meaning in metalwork' in Ursula Klein and Emma C. Spary (eds), *Materials and expertise in early modern Europe: Between market and laboratory* (Chicago, IL: Chicago University Press, 2010), pp. 29–49.

Smith, Pamela H. (ed.), *Entangled itineraries: Materials, practices, and knowledges across Eurasia* (Pittsburgh, PA: University of Pittsburgh Press, 2019).

Smith, Pamela H. and Benjamin Schmidt (eds), *Making knowledge in early modern Europe: Practices, objects, and texts, 1400–1800* (Chicago, IL: Chicago University Press, 2007).

Smith, Pamela H., Amy R. W. Meyers and Harold J. Cook (eds), *Ways of making and knowing: The material culture of empirical knowledge, 1400–1850* (Ann Arbor, MI: University of Michigan Press, 2014).

Snodin, Michael and John Styles (eds), *Design and the decorative arts, Britain 1500–1900* (London: V&A Publications, 2001).

Somers, Tim, *Ephemeral print culture in early modern England: Sociability, politics and collecting* (Martlesham: The Boydell Press, 2021).

Spufford, Margaret, *Small books and pleasant histories: Popular fiction and its readership in seventeenth-century England* (Cambridge: Cambridge University Press, 1981).

Steedman, Carolyn, *Labours lost: Domestic service and the making of modern England* (Cambridge: Cambridge University Press, 2009).

Steedman, Carolyn, *An everyday life of the English working class: Work, self and sociability in the early nineteenth century* (Cambridge: Cambridge University Press, 2013).

Stewart, Larry, 'Experimental spaces and the knowledge economy', *History of Science*, 45:2 (2007), pp. 155–77.

Stobart, Anne, *Household medicine in seventeenth-century England* (London: Bloomsbury, 2016).

Stobart, Jon and Mark Rothery, *Consumption and the country house* (Oxford: Oxford University Press, 2016).

Styles, John and Amanda Vickery (eds), *Gender, taste, and material culture in Britain and North America, 1700–1830* (New Haven, CT: Yale Center for British Art, 2006).

Sundberg Wall, Cynthia, *The prose of things: Transformations of description in the eighteenth century* (Chicago, IL: Chicago University Press, 2006).

Sweet, J. H., 'Mutual misunderstandings: gesture, gender, and healing in the African Portuguese world', *Past and Present*, 203:Supplement 4 (2009), pp. 128–43.

Taylor, Charles, *Modern social imaginaries* (Durham, NC: Duke University Press, 2003).

Terrall, Mary, *'Catching nature in the act': Réaumur and the practice of natural history in the eighteenth century* (Chicago, IL: University of Chicago Press, 2014).

Thirsk, Joan, *Economic policy and projects: The development of a consumer society in early modern England* (Oxford: Clarendon Press, 1978).

Thirsk, Joan, *Food in early modern England: Phases, fads, fashions 1500–1760* (London: Hambledon Continuum, 2006).

Thompson, Edward P., *The making of the English working class* (London: Victor Gollancz, 1963).

Thompson, Edward P., 'Time, work-discipline and industrial-capitalism', *Past & Present*, 38 (1967), pp. 56–97.

Tierney, James, 'Periodicals and the trade, 1695–1780' in Michael F. Suarez and Michael L. Turner (eds), *The Cambridge history of the book in Britain*, vol. 5, 1695–1830 (Cambridge: Cambridge University Press, 2009), pp. 479–97.

Tobin, Beth Fowkes, *The Duchess's shells: Natural history collecting in the age of Cook's voyages* (London: Yale University Press, 2014).

Topham, Jonathan R., 'Focus: Historicizing popular science, introduction', *Isis*, 100:2 (2009), pp. 310–68.

Towsey, Mark R. M., *Reading history in Britain and America, c.1750–c.1840* (Cambridge: Cambridge University Press, 2019).

Trentmann, Frank, *Empire of things: How we became a world of consumers, from the fifteenth century to the twenty-first* (London: Penguin, 2016).

Tronto, Joan, *Moral boundaries: A political argument for an ethic of care* (New York: Routledge, 1992).

Uglow, Jennifer, *The lunar men: The friends who made the future* (London: Faber, 2002).

Valleriani, Matteo (ed.), *The structures of practical knowledge* (Cham: Springer, 2017).

Vickery, Amanda, *The gentleman's daughter: Women's lives in Georgian England* (London: Yale University Press, 2003).

Vickery, Amanda, 'His and hers: Gender, consumption and household accounting in eighteenth-century England', *Past & Present*, 1:supplement 1 (2006), pp. 12–38.

Vickery, Amanda, *Behind closed doors: At home in Georgian England* (London: Yale University Press, 2010).

von Oertzen, Christine, Maria Rentetzi and Elizabeth S. Watkins, 'Finding science in surprising places: Gender and the geography of scientific knowledge. Introduction to "Beyond the academy: histories of gender and knowledge"', *Centaurus*, 55 (2013), pp. 73–80.

Warde, Alan, 'Consumption and theories of practice', *Journal of Consumer Culture*, 5:2 (2005), pp. 131–53.

Watts, Ruth, *Women in science: A social and cultural history* (Abingdon: Routledge, 2007).

Weatherill, Lorna M., *Consumer behaviour and material culture, 1660–1760* (London: Economic and Social Research Council, 1985).

Webb Jr., James L. A., 'The mid-eighteenth century gum Arabic trade and the British conquest of Saint-Louis du Sénégal, 1758', *The Journal of Imperial and Commonwealth History*, 25:1 (1997), pp. 37–58.

Werrett, Simon, 'The techniques of innovation: Historical configurations of art, science, and invention, from Galileo to GPS' in Dieter Daniels and Barbara U. Schmidt (eds), *Artists as inventors: Inventors as artists* (Stuttgart: Hatje Cantz, 2008), pp. 54–69.

Werrett, Simon, 'Recycling in early modern science', *British Journal for the History of Science*, 46:4 (2013), pp. 627–46.

Werrett, Simon, 'Household oeconomy and chemical inquiry' in Lissa L. Roberts and Simon Werrett (eds), *Compound histories: Materials, governance and production, 1760–1840* (Leiden: Brill, 2018), pp. 35–56.

Werrett, Simon, *Thrifty science: Making the most of materials in the history of experiment* (Chicago, IL: Chicago University Press, 2019).

White, Gilbert (A. Secord (ed.)), *The natural history of Selbourne* (Oxford: Oxford University Press, 2013).

White, Jonathan, 'Review essay: A world of goods? The "consumption turn" and eighteenth-century British history', *Cultural and Social History*, 3:1 (2006), pp. 93–104.

Whitley, Richard, 'Knowledge producers and knowledge acquirers: Popularisation as a relation between scientific fields and their publics' in Terry Shinn and Richard Whitley (eds), *Expository science: Forms and functions of popularisation* (Dordrecht: D. Reidel, 1985), pp. 3–28.

Whittaker, Katie, 'The culture of curiosity' in N. Jardine, J. A. Secord and E. C. Spary (eds), *Cultures of natural history* (Cambridge: Cambridge University Press, 1996), pp. 75–90.

Whittle, Jane, 'A critique of approaches to "domestic work": Women, work and the pre-industrial economy', *Past & Present*, 243 (2019), pp. 35–70.

Whittle, Jane and Mark Hailwood, 'The gender division of labour in early modern England', *The Economic History Review*, 73:1 (2020), pp. 3–32.

Whyman, Susan, *The pen and the people: English letter writers, 1660–1800* (Oxford: Oxford University Press, 2009).

Whyman, Susan, *The useful knowledge of William Hutton: Culture and industry in eighteenth-century Birmingham* (Oxford: Oxford University Press, 2018).

Williamson, Gillian, *British masculinity in the* Gentleman's Magazine, *1731–1815* (Basingstoke: Palgrave Macmillan, 2016).

Williamson, Gillian, *Lodgers, landlords, and landladies in Georgian London* (London: Bloomsbury, 2021).

Wilson, C. Anne, *Water for life: A history of wine, distilling and spirits, 500 BC to AD 2000* (Totnes: Prospect, 2006).

Wilson, C. Anne, 'Stillhouses and stillrooms' in Pamela A. Sambrook and Peter Brears (eds), *The country house kitchen, 1650–1900* (Stroud: The History Press, 2010), pp. 129–43.

Woods, C. J., 'Butler, Isaac', *Dictionary of Irish biography* (2009), https://doi.org/10.3318/dib.001249.v1.

Wrigley, E. A., *Energy and the English industrial revolution* (Cambridge: Cambridge University Press, 2010).

Yale, Elizabeth, 'Marginalia, commonplaces, and correspondence: Scribal exchange in early modern science', *Studies in History and Philosophy of Biological and Biomedical Sciences*, 42 (2011), pp. 193–202.

Yale, Elizabeth, 'Making lists: Social and material technologies for seventeenth-century British natural history' in Pamela H. Smith, Amy Meyers and Harold Cook (eds), *Ways of making and knowing: The material culture of empirical knowledge, 1400–1850* (Ann Arbor, MI: University of Michigan Press, 2014), pp. 280–301.

Yale, Elizabeth, *Sociable knowledge: Natural history and the nation in early modern Britain* (Philadelphia, PA: University of Pennsylvania Press, 2016).

Yale, Elizabeth, 'A letter is a paper house: Home, family, and natural knowledge' in Cala Bittel, Elaine Leong and Christine von Oertzen (eds), *Working with paper: Gendered practices in the history of knowledge* (Pittsburgh, PA: University of Pittsburgh Press, 2019), pp. 145–59.

Dissertations and theses

Avendal, Christel, 'Heightened everydayness: Young people in rural Sweden doing everyday life' (PhD thesis, Lund University, 2021).

Cowan, Steven, 'The growth of public literacy in eighteenth-century England' (PhD thesis, Institute of Education, University of London, 2012).

Day, Julie, 'Elite women's household management: Yorkshire 1680–1810' (PhD thesis, University of Leeds, 2007).

Lochrie, Eleanor, 'A study of lending libraries in eighteenth-century Britain' (MSc dissertation, University of Strathclyde, 2015).

Mather, Ruth, 'The home-making of the English working-class: Radical politics and domestic life in late Georgian England, *c.* 1790–1820' (PhD thesis, University of London, 2016).

McLoughlin, Riana, '"The sober duties of life": The domestic and religious lives of six Quaker women in Ireland and England, 1780–1820' (MA dissertation, University College Galway, 1993).

Osborn, Sally A., 'The role of domestic knowledge in an era of professionalisation: Eighteenth-century manuscript medical recipe collections' (PhD thesis, University of Roehampton, 2016).

Paskins, Mat, 'Sentimental industry: The Society of Arts and the encouragement of public useful knowledge, 1754–1848' (PhD thesis, University College London, 2014).

Stobart, Anne, 'The making of domestic medicine: Gender, self-help and therapeutic determination in household healthcare in south-west England in the late seventeenth century' (PhD thesis, Middlesex University, 2008).

Internet sources

British History Online: www.british-history.ac.uk/survey-london/vol2/pt1/pp61-64 (accessed 15 October 2021).

Buildings of Ireland, 'Strokestown Park': www.buildingsofireland.ie/building-of-the-month/strokestown-park-house-cloonradoon-td-strokestown-county-roscommon/ (accessed 2 August 2021).

Conroy, Rachel, 'Country house brewing' (13 March 2018): www.pressreader.com/ (accessed 25 February 2022).

'Cultures of knowledge': www.culturesofknowledge.org (accessed 26 July 2019).

The History of Parliament: www.historyofparliamentonline.org/volume/1820-1832/member/townley-richard-1786-1855 (accessed 2 August 2021).

Howes, Anton, 'The relevance of skills to innovation during the British industrial revolution, 1547–1851', working paper (2017): www.antonhowes.com/uploads/2/1/0/8/21082490/howes_innovator_skills_working_paper_may_2017.pdf (accessed 17 July 2021).

The Irish Aesthete, 'Strokestown Park': https://theirishaesthete.com/tag/strokestown-park/ (accessed 27 February 2022).

LUCK Lund Centre for the History of Knowledge: www.newhistoryofknowledge.com/2021/08/27/everyday-knowledge-as-unnoticed-knowledge/ (accessed 8 February 2022).

Shropshire History: http://search.shropshirehistory.org.uk/collections/getrecord/CCS_MSA271/ (accessed 12 June 2021).

Index

Note: 'n.' after a page reference indicates the number of a note on that page.